John Harvey Kellogg

The household manual of hygiene, food and diet, common diseases, accidents and emergencies

Useful hints and recipes

John Harvey Kellogg

The household manual of hygiene, food and diet, common diseases, accidents and emergencies
Useful hints and recipes

ISBN/EAN: 9783337201166

Printed in Europe, USA, Canada, Australia, Japan

Cover: Foto ©berggeist007 / pixelio.de

More available books at **www.hansebooks.com**

THE

HOUSEHOLD MANUAL

—OF—

HYGIENE, FOOD AND DIET, COMMON DISEASES,
ACCIDENTS AND EMERGENCIES, AND
USEFUL HINTS AND RECIPES.

By J. H. KELLOGG, M. D.

PUBLISHED AT
THE OFFICE OF THE HEALTH REFORMER,
BATTLE CREEK, MICH.
1877.

PREFACE.

As indicated by the title page, this little work deals with quite a variety of topics. It is thought, however, that all the subjects considered will be found usefully suggestive to every household. The aim has been to make the work eminently practical in character, and to condense into the smallest space the greatest possible amount of information.

The suggestions and hints given under the head of "Household Hygiene," if thoroughly appreciated and applied, will obviate a very large proportion of the ills and suffering incident to domestic life.

The section on "Food and Diet" contains much which may be new to a majority of those who have never investigated the subject from the standpoint of health. It is not intended to be in any sense complete, the object being only to call attention to a few of the ways in which disease and premature death are occasioned by errors in diet. Those who are interested to pursue the subject further should send to the Office of publication for other works treating more at length.

In "Simple Remedies for Common Diseases" are given directions for treating many common maladies with such remedies (with few exceptions) as are to be found in any household.

"Accidents and Emergencies" will be found to afford such information as may enable a person to be the means of saving many lives if it is carefully and promptly applied at the proper time.

CONTENTS.

———·———

HOUSEHOLD HYGIENE.

FOOD AND DIET.

SIMPLE REMEDIES FOR COMMON DISEASES.

HOUSEHOLD HYGIENE.

"HEALTH is wealth" is a trite maxim, the truth of which every one appreciates best after having suffered from disease. Indeed, health is a most priceless treasure. When deprived of it, we are willing to exchange for it everything else we possess; yet when well, we squander it ruthlessly, disregarding the plainest rules of health, regardless of consequences. It is only when sick, and suffering the result of transgression of Nature's laws, that we begin to appreciate the value of health, and the importance of regarding carefully the conditions upon which health depends.

State and National Health Boards and Committees certainly do excellent work for communities and nations; but the real influence which they exercise over the health of individuals is insignificant when compared with that which may be, and indeed is, exercised by the matrons of the various households which make up villages, cities, and nations. City authorities may exercise a rigid surveillance over all the avenues through which disease is known to enter; they may keep the public streets cleanly, introduce costly means of

supplying water, and cause the removal beyond
the suburbs of slaughter-houses, tanneries, soap-
boiling establishments, and noisome chemical
works; but if the seeds of death and disease are
allowed to germinate and flourish in each separate dwelling, and around each fireside, what
favorable results can be expected?

All reforms must begin at home, to be effective; and we would urge upon all parents the
importance of careful attention to the simple suggestions which are herein offered, by means of
which they may be able to save themselves and
their families from numerous illnesses, with their
attendant inconveniences, expense, and suffering.

Fresh Air.—From the first quick gasp of in-
fancy to the last feeble sigh of old age, the prime
necessity of life is air. Air is food for the lungs,
as bread is food for the stomach. Millions more
people die from want of lung food than from a de-
ficiency of other aliment. The Creator has provided the necessary article in generous abundance, fresh, pure, and free to all. If we do not
get enough, it is our own fault, for when we close
our doors and windows the closest, this vitaliz-
ing, invigorating element is whizzing and howl-
ing close around outside, seeking to find an entrance.

People who nail up their windows, stop every
crack and crevice in the walls, line the door casing with felt, and fix a patent thing under the
door as a sort of air-trap to catch the occasional

whiffs of pure air which might otherwise get in, are barricading themselves against their best friend. A man who should so studiously and deliberately deprive himself of the means of procuring ordinary food, would be pronounced a suicide. Is he any less a transgressor—though ignorantly so—who deprives himself and his family of a still greater necessity, pure air?

The demand for pure air is the most imperative of all the wants of the system. When deprived of air, an individual will die sooner than from deprivation of any other of the essentials of life. A person may live several weeks without solid food of any kind, several such cases having been noted by eminent authorities. When deprived entirely of drink, life sooner becomes extinct. But if an individual be deprived of air, death occurs in a few minutes.

Sources of Impure Air.—The sources from which the air may become contaminated are so very numerous that we cannot dwell at length upon all of them in so concise a treatise as this. We can only notice some of the more common.

Poisonous Gases.—Of the numerous poisonous gases which mingle with the air we breathe, *carbonic acid*, or, more properly, *carbon di-oxide*, is the most common and abundant of all. This gas is heavier than air, and, consequently, it collects in such low places as deep wells, old cellars, caves, and deep valleys. It is produced by combustion and decay in vast quantities, and would soon accumulate to a fatal extent were it not for the fact that while it is a fatal poison to man, it constitutes a necessary food for plants.

One important fact to be remembered respecting the properties of this gas is its want of odor when pure, so that its presence cannot always be detected by the sense of smell as can most poisonous gases.

In Italy there is a curious cave, the bottom of which is covered with carbon di-oxide to a depth of about two feet. Travelers can explore the cavern with perfect impunity; but dogs or other small animals which accompany them, are quickly suffocated.

This gas is produced in great volumes in the burning of lime, being driven off by the excessive heat. Cases of poisoning by this gas have occurred, in which persons have lain down to sleep beside the warm kiln and have been suffocated by the escaping gas.

Amount of Carbonic-Acid Gas Produced.— This gas is formed within the body, and finds its way out through the lungs. An adult man produces about seven gallons of the gas per hour. A gas-light produces several times as much. An ordinary candle produces quite a considerable quantity. Large quantities are produced in a stove or fireplace; but that which is generated in this manner is usually carried away with the smoke, and does not escape into the room.

Carbonous Oxide is an exceedingly poisonous gas which is formed by imperfect oxidation of .the fuel, which is frequently the result of deficient draught. The gas is often found in air-tight stoves furnished with close dampers. One remarkable property of the gas is its penetrating power. It will pass directly through cast-iron, especially when it is heated. A few years ago a whole school were poisoned by this gas, several nearly to a fatal extent. It paralyzes the blood corpuscles, and thus renders respiration impossible. It is a much more poisonous gas than carbonic acid, and is fatal in much more minute doses. In the case of the school referred to, the teacher had turned the damper of the stove so as to cut off the draught while the stove was hot, and in a short time discovered that a large share of the students were falling into a state of stupefaction. This is a good illustration of the importance of always leaving sufficient draught to carry off the products of combustion. As this gas, like carbonic-acid gas, has no odor, it will only be detected by its effects.

Sulphureted Hydrogen is a still more poisonous gas which frequently finds its way into the air which human beings breathe. Fortunately it has a very bad smell, the characteristic odor of rotten eggs, in which it is always present. This gas is developed wherever animal matter is undergoing decomposition. It is poured forth in volumes from cess-pools, sewers, gutters, drains, privy vaults, neglected cellars and cisterns, and every other place where animal substances are allowed to putrefy. It is this gas which gives to most decaying substances their offensive character. In the gutters of back streets and alleys in our large cities, this gas is sometimes produced in such enormous quantities that its active chemical properties become very perceptible, as will be shown by the following anecdote related by a professor of chemistry in one of our State Universities :—

"A young lady who was entirely innocent of any knowledge of chemistry or chemical facts, emerged from an elegant mansion in New York City, fully equipped for an afternoon promenade, with face artistically painted *a la mode*. Her course, unfortunately, lay for a little distance through a portion of the city where the drainage was imperfect, and the air was consequently redolent with that wonderfully pungent and active gas which is so characteristic of rotten eggs— sulphureted hydrogen. Of course the lady could not be unconscious of the presence of some noxious element in the atmosphere; but she was nevertheless wholly ignorant of its chemical properties. Her ignorance did not, however, deter the gas from manifesting its most vigorous affinities for the lead paint upon her cheeks, of which she had abundant evidence as she stood before a mirror, upon her return home, and viewed the swarthy appearance of her countenance, which would have been very becoming to a representative member of the African race."

Ammonia, Sulphurous Oxide, with various other noxious gases, find their way into the air in numerous ways, and exert a deleterious influence upon the health.

Germs.—Some of the most active and powerful enemies of human life are those which are the most insignificant in size, and hence the most likely to escape detection. Wherever decay of either animal or vegetable matter is taking place, myriads of microscopic plants flourish in great luxuriance. These numerous species of fungi are generated by spores which float about in the air, and, finding lodgment in favorable places, develop in plants which, in turn, produce countless numbers of other spores which quickly find their way into the air to repeat the same process elsewhere.

It is the presence of these little germs which causes the fermentation of yeast and the "rising" of bread, together with the "working" of wine and cider, the "spoiling" of canned fruits and other preserved products, the souring of milk, and all kinds of decay and decomposition.

The conditions required for the growth and development of these minute organisms are warmth and moisture. In the winter they are paralyzed by the cold; but so soon as the vernal sun appears, they spring quickly into life and activity. As before remarked, these little living particles fill the air. Sometimes, and in some places, the air is heavily laden with them ; again, they are present in much more limited numbers. They are, of course, taken into the lungs with the air which is breathed, and thus they find entrance into the system, and under certain circumstances produce dangerous and fatal diseases. Beware of germs !

Dust.—It is next to impossible to obtain air wholly free from dust. Its constant motion lifts and holds suspended little particles of various substances which are more or less injurious to health, unless the quantity is very small indeed. Some trades, as stone-cutting, coal-heaving, rag-picking, cotton and wool spinning and weaving, and other avocations which involve the production of considerable quantities of dust, expose the workmen to an atmosphere loaded with fine particles which are drawn into the lungs with every breath, and, finding lodgment there, may induce irritation and still more serious disease of those organs. By a wonderful provision of nature, the finer particles of dust, if in small quantity, may be wholly removed so that they will not pass down into the more delicate air-cells of the lungs. But if the quantity of dust is great, this provision fails to afford protection.

The inhalation of dust is one of the causes of consumption. Post-mortem examination of the lungs of persons who had died from this cause showed the lungs to have acquired the color of the particles inhaled; and, in some cases, they contained so large a quantity of sand that they felt gritty to the touch.

Great care should always be taken.to avoid dust as much as possible. In sweeping carpets and dirty floors, a person is exposed to injury unless some precaution, such as sprinkling the floor or moistening the broom, is taken to prevent filling the air with dirt. There are very few people who would not turn with disgust from food which was filled with particles of coal or sand, covered with dust, and gritty to the teeth. Yet the same persons will take their gaseous food in precisely the same condition without remonstrance.

Organic Poison.—Gases, germs, and dust are most prolific sources of disease and death which attack man from the air; but there is yet another enemy of life more potent still, which lurks, too often unsuspected, in the air we breathe. Very little, indeed, is known of the real nature of this poison, since it has, in considerable degree, eluded the efforts of the chemist to submit it to analysis; but it is of organic origin, and hence is known by the term *organic poison.* This poisonous element is introduced into the air chiefly by means of respiration, together with exhalations from the skin. It is one of the most noxious poisons ever present in air. It will produce death much sooner than most other impurities found in the air. Experiments upon animals have shown that a mouse will die in a few minutes when confined in air heavily charged with this poison.

The moisture which condenses on the inside of the windows of an occupied room in a cold day contains the poison in solution. If a little is collected in a vial and set away, it will soon become intensely fetid and offensive. It is this poison which gives to an unventilated room the close, fusty odor with which every one is familiar. One who has been long in the room will not observe it; but it is very distinct to a person coming in directly from the pure air outside.

Malaria.—The great curse of large areas of the most beautiful portions of our country is malaria. With reference to the exact nature of this cause of disease there has been a great amount of discussion. The most plausible theory is that advanced by Dr. Salisbury, of the Ohio Medical College, who claims to have demonstrated that the exciting cause of malarial disease is the spores of a certain species of fungi. According to this authority, the ague-plant flourishes in low grounds which are frequently submerged, but are covered with water but a portion of the time. It is well known that malarial diseases, as ague or intermittent fever, remittent or bilious fever, and typho-malarial fever, are most prevalent in just such localities as are favorable for the production of the so-called ague-plant. An unusually dry season is almost certain to be followed by an unusual number of cases of remittent fever and ague in the vicinity of marshes, mill-ponds, and shallow streams, the beds of which are exposed during the drouth.

The malarial miasm is often carried several miles from its source, so that immediate proximity to the latter is not necessary for contraction of malarial disease. Nevertheless, the observance of a few precautions will greatly lessen the liability to the disease. The following hints will be found of service :—

1. Avoid the vicinity of malarious districts during the evening and early morning, as the malaria settles nearer the surface of the ground at those times.

2. Secure, if possible, a dense growth of trees between a malarious district and the residence, as the foliage of trees affords a very efficient barrier to the miasm.

3. In case the above is impracticable, the same purpose may be accomplished, in considerable degree, at least, by planting between the house and the source of malaria a large area of sunflowers, which are said to possess the power to destroy the malarial poison by the production of ozone.

4. The liability to the disease may also be very greatly lessened by keeping the system in as free a condition as possible by avoiding such habits and such articles of food as will impair the function of the liver, skin, kidneys, and other depu-

rating organs. By this means the system may be enabled to eliminate the poison without occasioning disease.

How to Ventilate.—The only way to get fresh air is to obtain it from out-of-doors, by exchanging the foul air within for pure air without.

How much fresh air do we need? Every man needs enough to dilute the poison which he exhales sufficiently to render it harmless. To effect this, a quantity of air 5,000 times as great as the amount of carbon di-oxide produced, is required. In other words, 5,000 gallons of pure air are necessary to render harmless one gallon of carbon di-oxide. A man produces a gallon of this poison every twelve minutes, or five gallons an hour; hence, he requires 5,000 gallons of pure air every twelve minutes, or 25,000 gallons each hour—more than 3,000 cubic feet.

To ventilate well, there must be two openings; one at the bottom, and the other at the top of a room. What! shall we open the windows at top and bottom on a cold, wintry day? Certainly. Cold air is not poison. Plenty of air and a rousing fire are cheaper in the long run than foul air and less fire.

But will not cold air produce colds, and lung fevers, and pleurisies, and consumptions? People don't catch cold in open sleighs nor when walking in the wind. Draughts of cold air upon a small portion of the body only, will occasion cold; but there need be no draughts. Avoid them in this way :—

Make a strip of board, three or four inches wide, just the length of the window casing. Fit it beneath the lower sash. This makes an opening between the two sashes where they overlap. Here the air can enter, and being thrown upward toward the ceiling, it will be productive of no harm to any one.

Another way : Lower a window at the top on one side of the room, and on the opposite side raise another a little at the bottom. Place a screen of fine netting in front of each, and the room will be pretty well ventilated without draughts.

Unless a strong wind is blowing, the window should be lowered one inch for each occupant of the room. A window should be raised an equal amount upon the opposite side to allow a circulation of the air.

The old-fashioned fire-place was a most efficient ventilator. It is a good omen that fire-places are again coming into use. The most fashionable parlors in the large cities are now heated by them.

If flues are used in ventilating rooms, it is absolutely necessary that the air in them should be heated several degrees higher than that in the rooms, to secure a draught. There should be two openings into the flue; one near the ceiling to be used when necessary to change the air rapidly, and the other at the floor to be open constantly.

Never sleep in a room which is unventilated.

To Destroy Foul Odors. — Abundance of fresh air is the best deodorizer. There is no substitute for ventilation. Pure air washes away foul smells as water washes away dirt. One removes solid filth, the other gaseous filth. If the offensive body is movable, be sure to remove it. If not, apply something to destroy it. Several agents will effect this.

If it can be safely done, set fire to the foul mass; or, if this is undesirable, heat it almost to the burning point.

Apply very dry, finely pulverized earth. Clay is the best material. Finely powdered charcoal which has been freshly burned, is quite as good as earth. Dry coal or wood ashes are excellent.

Make a solution of permanganate of potash, dissolving one ounce in a gallon of water. Add this to the offensive solid or fluid until it is colored like the solution. This is an excellent deodorizer. It is needed in every household. A supply of the solution should be kept constantly on hand, ready for use.

Copperas dissolved in water in proportion of one pound to the gallon of water is cheaper, and may be used when large quantities are needed. Apply it freely.

Bromo-chloralum is a very good deodorizing agent, but is rather expensive.

Chlorine gas, chloride of lime, ozone, and numerous other agents, are effective when rightly used.

Disinfecting Fluid.—The following is a recipe for one of the cheapest and most efficient disinfecting fluids known :—

Heat two pounds of copperas in an old kettle for half an hour, stirring frequently. When cold, dissolve the copperas in two gallons of water. Add two ounces of carbolic acid, and mix well together. A pint of this solution poured into the kitcken sink every few days will keep it free from odors. It will also be found very useful for disinfecting the discharges of typhoid-fever patients, for which purpose a little should be kept in the vessel constantly. Even privy vaults can be kept in a comparatively harmless condition by the liberal use of this solution.

Cess-pools.—Drains, sewers, and cess-pools, connected with a house, are often sources of serious disease. The kitchen sink is not unfrequently the door through which the germs of disease silently creep into a household and develop into disease and death, the cause of which remains a mystery, and is attributed to the inscrutable dealings of Providence.

In the summer, draughts are produced in the room, which suck up the filthy gases which are formed in the cess-pool or sewer, through the drain-pipe—unless it is furnished with an efficient water-trap, which is not usually the case.

In the winter, the gases of the cess-pool are naturally warmer than the air above, and so they rise and find their way into the house, filling it with invisible poison, which is breathed, and thus taken into the blood, by every occupant of the dwelling. Thousands of valuable lives are annually sacrificed in this way.

How shall this evil be remedied? In cities, the problem is a difficult one, unless sewers can be replaced by the dry-earth system. In the country and in small towns, it is easily cured thus :—

Make the cess-pool some little distance from the house. Place in communication with it a wooden ventilating flue sixteen or eighteen feet in height, and four to six inches square inside. This will carry off the foul gases under ordinary circumstances, but it will sometimes be found inefficient; hence, a water-trap should be formed in the drain-pipe, just beneath the sink, by bending the pipe so that it will retain constantly three or four inches of water.

A still better way is to connect the drain-pipe with the chimney or stove-pipe, by means of a pipe of suitable size. This will secure ventilation of the drain ; and if the connecting pipe joins the drain-pipe just beneath the sink, the protection will be perfect. All joints should be airtight, and the outlet from the sink should be plugged tightly when there is no fire in stoves communicating with the chimney.

Another valuable precaution is this : Pour into the sink, two or three times a week, a gallon of water in which a pound of copperas has been dissolved. A few crystals of copperas kept constantly in the sink could do no harm. It is very cheap when bought by the quantity.

A new cess-pool should be made at least once a year, or the old one should be thoroughly cleaned.

Under the House.—Many families who wonder why "some of the children are sick all of the time," can find the cause underneath the floor. Nearly all houses have cellars. Here are stored all sorts of things for winter use—dead things and live things, articles to eat and fuel to burn, old boxes and barrels, heaps of coal, bins of vegetables, etc., etc. The coal and wood are continually sending up foul gases and germs. Many of the vegetables undergo decay, and add greatly to the production of disease elements.

Besides the cellar there is usually an open space under the other portions of the house, between the foundation walls. This space is large enough to admit chickens, dogs, cats, rats, even pigs, and other small animals, but not sufficiently large to allow room for clearing it. Here various small animals find a hiding-place, and often die. Being out of sight and reach, they are not discovered even when the stench of their decaying bodies becomes distinctly manifest.

All the foul gases engendered in these various ways pass upward into the house, filling every room, condensing in fetid moisture upon the walls, and poisoning all who breathe in the house. What shall be done?

Cellars under a house are rather prejudicial to health, even at best. As they are commonly used, they are very greatly so. If there must be cellars beneath the house, they should be large, light, and well ventilated. Every week, at least, the cellar windows should be opened wide to allow free change of air. A good way to ventilate a cellar is to extend from it a pipe to the kitchen chimney. The draught in the chimney will carry away the gases which would otherwise find their way into the rooms above.

Cellars should be kept clear of decaying vegetables, wood, wet coal, and mold. The walls should be frequently whitewashed, or washed with a strong solution of copperas. The importance of some of these simple measures cannot well be overestimated.

Houses should be built so high above the ground that the space beneath can be easily cleared every few months.

Moldy Walls.—Many people who do not appreciate the importance of sunshine as they should, allow mold and mildew to accumulate upon their walls in damp weather, especially in nooks and corners that will be unobserved, never thinking that any harm will come from so doing. Such are ignorant of the fact that each patch of mold is a forest of myriads of little plants which are constantly throwing off into the air myriads of germs to be inhaled by the occupants of the house. In ancient times, collections of fungi of this sort were looked upon as matters of such serious import as to render a house wholly unfit for habitation until it had been thoroughly cleansed. A house with moldy walls was said to be affected with the plague of leprosy, and if the discolored, moldy spots recurred after having been thoroughly cleansed away, the house was abandoned and torn down. No one was allowed to occupy it unless every trace of the mold could be wholly removed. A tithe of the same care now would save thousands of deaths annually.

Privies.—As ordinarily constructed and managed, these necessary institutions are most prolific sources of disease. The animal excretions which are left to accumulate in them undergo still further putrefactive changes, which result in the development of the most pestilential germs and gases. Here is where the terrible typhoid poison originates. Deep vaults should never be allowed under any circumstances.

The best way to manage a privy is this: Early in the spring fill up the old vault, if there is one, even with the surface. Raise the building a little. Have made at the tin-shop a sufficient number of pans of thick sheet-iron. The pans should be about two feet square, and two inches and a half deep. Each should be furnished with a long bail, and a strong handle at one side about a foot in length. In using these pans, fill each half full of fine, dry dirt—not sand—or ashes, and shove it into position, allowing the bail to fall back upon the handle behind. By the addition of a little dry dirt several times a day, all foul odors will be prevented. The contents of the pans ought to be removed every night in the warmest weather of summer, the pans being replaced with a fresh supply of dry earth. During cooler weather, if little used, the pans will require emptying but once a week, if they are kept well supplied with dry earth. The contents of the pans may be buried or removed to a proper place at a distance from any dwelling.

For convenience, it is found to be an excellent plan to hire a scavenger to attend to the pans at regular, stated times. Fifteen or twenty in a community can unite on the same plan, and thus make the expense very slight for each.

About the first of December, the pans may be removed and a shallow vault dug. The vault should not exceed two feet in depth, and it should not be tightly inclosed. This will allow the contents of the vault to freeze. They may be removed several times during the winter, and should be kept covered with dry dirt, which should be procured in sufficient quantity in the fall.

Sunshine.—In caves, mines, and other places which are excluded from the light, plants do not grow, or, at most, they attain only a sickly development. The same is true of animals. In the deep valleys among the Alps of Switzerland, the sun shines only a few hours each day. In consequence, the inhabitants suffer terribly from scrofula and other diseases indicative of poor nutrition. The women, almost without exception, are deformed by huge goiters, which hang pendant from their necks unless suspended by a sling. A considerable portion of the males are idiots. Higher up on the sides of the mountains, the inhabitants are remarkably hardy, and are well developed, physically and mentally. The only difference in their modes of life is the greater amount of sunshine higher up the mountain side. When the poor unfortunates below are carried up the mountain, they rapidly improve.

Throw open the blinds and draw aside the window curtains. Never mind if the carpets do fade a little sooner. The pale cheeks will acquire a deeper hue, and the sallow skins will become of a more healthy color.

A sitting-room ought to be on the east or south side of a house, so that sunlight will be plentiful. House plants will not thrive in a north room. Women and children, who live mostly in the house, thrive no better in such a situation than plants. Sleeping-rooms should be aired and sunned every day.

House Plants in Sleeping-Rooms.—The supposition that house plants are injurious in sleeping and sick-rooms is a popular error. It is commonly supposed that plants draw the vitality of the patient, or poison the atmosphere in some way. This is wholly an error, if we except a few of the more strongly scented plants, which emit a somewhat poisonous odor, or which might in some cases be unpleasant to the senses of a nervous patient. Plants cannot draw vitality from animals. Indeed, they are the one great means which make human life possible; for if they did not purify the air, all animals would quickly perish.

Plants inhale carbon di-oxide during the day, and exhale oxygen. During the night, they inhale carbon di-oxide the same as in the daytime, but exhale a part of it again, along with the oxygen. They purify the air, then, during the night, but less than during the day.

A mouse and a growing plant can live together in an air-tight box. Alone, either one would die; together, they both thrive. Plants purify the air for human beings as well as for mice.

Plants also remove impurities from the air by means of the *ozone* which they produce, which is one of the most powerful disinfectants known. The laurel, hyacinth, mint, mignonette, lemontree, and fever-few are among the best ozone-producing flowers.

The cheerful aspect which flowers give to a

room, and the pleasant recreation which their care affords, are not the least of the advantages to be derived from them.

Beds and Bedding.—A cold, damp, musty bed has cost the world many a valuable life. The "spare bed" is a genuine terror to traveling ministers, and school-teachers who board around. A night spent in one of them is a certain cause of cold, headache, sore lungs, sore muscles, and stiff joints the next day. Never sleep in a room which has been unused for weeks, unaired, unwarmed, and secluded from sunlight, until the bedding, at least, has been thoroughly aired and dried, and the air of the room thoroughly changed by ventilation. Never offer such a room for the accommodation of a guest without treating it in the same way, unless it is desired to make him sick.

Feather-beds are very unhealthful. They not only undergo a slow decomposition themselves, thus evolving foul and poisonous gases, but they absorb the fetid exhalations from the body which are thrown off during sleep. By constant absorption, the accumulation soon becomes very great, and the feather-bed becomes a hot-bed of disease. Hair, cotton, straw, or husk mattresses are greatly superior to feathers from the standpoint of health.

Don't cling to the old feather-bed because it is an heir-loom. The older it is, the worse it is. Only think of the amount of diseased germs which must be stowed away in a sack of feathers which has done service during a hundred years or more! Subject to all the accidents and emergencies of domestic life it has, perhaps, carried a half-dozen patients through typhoid fever, and pillowed the last months of the gradual dissolution of a consumptive, besides being in constant use the balance of the time.

Barnyards.—The close proximity of barnyards, hen-coops, and hog-pens to human dwellings is a frequent cause of serious and fatal disease. The germs which are developed in the filth abounding in those places, together with the noxious gases constantly arising from the decomposing excreta, are productive of disease when received into the system. Often, indeed, the well from which the family supply of water is obtained, will be located only a few feet from a reeking barnyard, or, as we have more than once seen, the well will, for convenience, be located within the yard itself. In consequence of the proximity, the water of the well will be contaminated by the soluble filth which percolates down through the porous earth and finds its way into the underground veins of water by which the well is fed.

Notwithstanding all these dangers, there are people who, incredible as it may seem, still hold to the absurd idea generated in the Dark Ages, when the streets of every city were one immense reeking cess-pool, that foul smells originating in the filthy ordure of horses and cows possess some healing properties. Not long ago when we appealed to a man to clear his barnyard, which had become a positive nuisance, being not more than half a dozen feet from the threshold of a dwelling-house, he retorted that he had always been informed, and as he thought by good authority, that a barnyard smell was the "healthiest kind of a smell," and was "especially good for consumptives." If there is such an absurd error prevalent, it ought certainly to be corrected. No foul, noxious odor can be of any possible advantage to the health. Barnyards should be located at least forty or fifty rods away from the dwelling, and wells should be located nearly as far removed from such sources of poisoning, to insure against water contamination, which is one of the most common causes of typhoid fever.

Cleansing Sick-Rooms.—A room which has been long occupied by a person suffering from chronic disease, or by a fever patient, or a case of small-pox or other contagious disease, ought to be very thoroughly cleansed before being occupied by others. The means by which this may be most efficiently done are these :—

1. Take out the windows, and give the greatest possible freedom to ventilation.

2. Remove the old paper from the walls, and burn it. Wash the bare walls with a solution of copperas, and then apply whitewash to the ceil-

ing. Cleanse the wood-work with a solution of chloride of lime.

3. Remove the carpet from the floor, the bedding from the beds, and every other fabric in the room, and thoroughly disinfect them before replacing.

4. If still more thorough disinfection is desired, remove from the room such furniture as will be injured by corrosive gases, close the windows tightly, and place in the center of the room a shallow stone or earthen vessel containing the following mixture : 4 oz. each of salt and black oxide of manganese, 3 fl. oz. of water, and $3\frac{1}{2}$ fl. oz. of sulphuric acid, or oil of vitriol. Mix the acid and the water first, let it cool, and then add it to the salt and oxide of manganese, which should be previously intimately mixed in the earthen vessel. Stir well with a stick, and then close the room as tightly as possible, stopping up the crevices. Chlorine gas will be slowly formed by this means, and it will destroy whatever organic matter there may be in the room. It will even penetrate the plaster on the walls.

In two or three days the room should be opened and thoroughly ventilated.

Disinfecting Clothing.—Clothing which has been exposed to contamination by contagion, if of little value, should be destroyed. If more valuable, it may be disinfected in any one of several ways.

1. Heat in an oven as hot as possible without

scorching, for an hour or two. A temperature of 250° will do no harm.

2. If the clothing is uncolored, or colored with mineral dyes, soak a few minutes in a solution of fresh chloride of lime of the strength of one pound of the chloride to a pailful of water. Afterward boil.

3. Soak for half an hour in boiling water to which carbolic acid has been added in proportion of an ounce to the gallon of water. Boil again in pure soft water, to remove the smell of the acid.

4. Expose for several hours in a close box to the fumes of burning sulphur. Air thoroughly afterward and wash.

Sick-Room Disinfection.—In such diseases as typhoid fever, dysentery, cholera, yellow fever, and diarrhea, the bowel discharges should be instantly disinfected and then removed as soon as possible. To do this readily and promptly, a strong solution of permanganate of potash or copperas should be kept constantly in the chamber vessel. Large vessels of water kept in the room and daily changed will absorb much of the gaseous poison. Carbolic acid, chloride of lime, and other odorous disinfectants, are offensive to the patient, and should not be used. Most thorough ventilation should be secured constantly. A little management will protect the patient from cold draughts, and there will be no danger of exposing him to cold, if he has the care of an attentive nurse, even if the ventilation is the most thorough.

House-Cleaning.—The semi-annual house-cleaning, although not a pleasant experience, is just as necessary as the original building of the house. Some important things are often overlooked in the general hurry and confusion.

The closets, garrets, clothes-rooms, stairways, and similar places need thorough renovation as well as more conspicuous rooms. The steam and gases from the kitchen find their way into all parts of the house, and are absorbed by the porous walls, or condense upon the wood-work. If not removed, they become sources of disease. The spare bedroom and parlor must not be neglected on account of having been little used, for the same reason.

Wood-boxes are too often neglected until the rubbish at the bottom becomes exceedingly foul, and occupies so much space that there is little room for anything else. Wet, souring, fermenting bark and chips, decaying apple cores, moldy leather, and similar elements which usually occupy a considerable portion of wood-boxes, contribute largely to the production of many febrile diseases.

New wall-paper should never be put on over old. The fresh paste, by its moisture, causes the fermentation of the old paste and the production of foul gases from the colors of the paper and the impurities which have been absorbed. If the old paper contained arsenic, the danger is increased tenfold, as arseniureted hydrogen is formed, one of the most fatal gases known. House-cleaning is one of the most important parts of domestic labor, and should not be trusted wholly to ignorant servants. It should be done under the constant supervision of an intelligent and thorough-going person. A little neglect to examine and thoroughly cleanse every nook and corner may result in the sacrifice of a human life. Too much importance cannot be attached to the necessity of care and painstaking in this matter.

Every dwelling should be thoroughly cleansed at least twice a year. Old carpets with their accumulated dust should be taken up and thoroughly beaten and cleansed, bed-ticks should be refilled if straw is used, every bed should be carefully examined for vermin, and a general renovation should take place.

Poisonous Paper.—Many cases of poisoning, some of which were fatal, have been traced to the arsenic contained in several of the colors of wall-paper. The most dangerous color is green. It is almost impossible to find a green paper which does not contain arsenic. Green window curtains are especially dangerous. The green dust which can be rubbed off from them is deadly poison. In rolling and unrolling the curtain it is thrown into the air and is breathed. The same poison is brushed off the surface of arsen-ical wall-paper into the air by the rubbing of pictures, garments, etc., which come in contact with it.

It is very easy to test papers of this kind be-fore buying, and it would be wise always to take this precaution. Take a piece of the paper and pour upon it strong aqua ammonia over a saucer. If there is any arsenic present, this will dissolve it. Collect the liquid in a vial or tube, and drop in a crystal of nitrate of silver. If there is arsenic present, little yellow crystals will make their appearance about the nitrate of silver. Arsenical green, when washed with aqua ammonia, either changes to blue, or fades.

Poisonous Aniline Colors. — Red flannel, stockings, and hat linings, and the striped stock-ings which have recently become fashionable, have occasioned serious poisoning in numerous cases. The aniline dyes with which they are colored are used in connection with arsenic, which is not wholly removed by the manufact-urers.

Hair Dyes and Cosmetics.—Any number of "Hair Dyes," "Hair Vigors," "Hair Renewers," "Hair Tonics," and various other compounds for application to the hair with the object of restor-ing its color or promoting its growth, have been invented during the last ten years. Many of these mixtures claim to be purely vegetable, and harmless. This is untrue of any of them. They contain, almost without exception, a very large amount of mineral poison. Lead, silver, and sulphur are the most common ingredients. The effects of applying such articles to the head are very serious. A few of the more prominent are the following :—

Headache, vertigo, irritation of the scalp, apoplexy, congestion of the brain, nervousness, sleeplessness, paralysis, and insanity. Numerous instances of all of these maladies have occurred as the result of using " hair dyes."

Gray hair is no disgrace. The healthful growth of the hair can be promoted by daily friction with cool soft water much better than by any quack lotion.

Cosmetics are equally dangerous. We have seen hopeless paralysis of the extensor muscles of the fore-arm, causing wrist-drop, produced by the use of paints for improving the complexion. Young ladies have destroyed their usefulness for life by this foolish practice. Lead colic is another result of the use of paints, many of which contain lead. Beware of them.

Hygiene of the Eyes.—These, the most delicate of the organs of sense, are often ruined by abuse. With good usage they will "last a lifetime." It is necessary to observe the following rules, to preserve the health of the eyes :—

1. Never use the eyes when they are tired or

painful, nor with an insufficient or a dazzling light. Lamps should be shaded.

2. The light should fall upon the object viewed from over the left shoulder, if possible; it should never come from in front.

3. The room should be moderately cool, and the feet should be warm. There should be nothing tight about the neck.

4. Hold the object squarely before the eyes, and at just the proper distance. Holding it too near produces near-sightedness. Fifteen inches is the usual distance.

5. Never read on the cars, when riding in a wagon or street-car, or when lying down. Serious disease is produced by these practices.

6. Do not use the eyes for any delicate work, reading, or writing, by candlelight, before breakfast.

7. Avoid using the eyes in reading when just recovering from illness.

8. Never play tricks with the eyes, as squinting or rolling them.

9. If the eyes are near-sighted or far-sighted, procure proper glasses at once. If common print must be held nearer than fifteen inches to the eye for distinct vision, the person is near-sighted. If it is required to be held two or three feet from the eye for clear sight, the person is far-sighted.

10. A near-sighted person should not read with

the glasses which enable him to see distant objects clearly.

11. Colored glasses (blue are the best) may be worn when the eye is pained by snow or sunlight, or by a dazzling fire or lamplight. Avoid their continued use.

12. Never patronize traveling vendors of spectacles.

13. Rest the eyes at short intervals when severely taxing them, exercising the lungs vigorously at the same time.

Keep Clean.—The skin, the superficial covering of the whole body, everywhere abounds in little mouths, or openings, called pores. There are more than 2,000,000 of these openings upon the surface of the body. Each one is the external orifice of a capillary tube which acts as a kind of sewer to convey away dead, effete, and decomposing matter from the body. Each of these purifying organs is constantly at work unless its mouth gets obstructed in some way. They are especially active in the summer season when the weather is warm, pouring out large quantities of perspiration in which the offensive matters are held in solution.

Now let us see what takes place if we pay no attention to the natural clothing with which we have been kindly provided. The sweat or insensible perspiration, with a load of impurities, is poured out of 2,000,000 little sewers, upon the surface of the body. The watery portion evaporates, leaving behind all the foul matter which it contained, which adheres to the skin. This is what occurs the first day. The next day an equal quantity is deposited in the same way, making, with the previous deposit, a thin film of dirt covering the skin. The third day the quantity has augmented to the consistency of varnish. The fourth day the person becomes completely encased in a quadruple layer of organic filth. By the fifth day, fermentation begins, and an unsavory and pungent odor is developed. The sixth day adds new material to the accumulating pollution, and still further increases the intensity of the escaping effluvia. Upon the seventh day a climax of dirtiness is reached. The penetrating, pungent fetor becomes intolerable. The person feels as though he had been bathed in mucilage or molasses. When he approaches his more cleanly friends, they look around to see if there is not some fragment of carrion adhering to his boot. But the individual himself is unconscious of any unpleasant odor, his nose having become accustomed to the stench; or if he recognizes it, he flatters himself that as no one can *see* the condition of his cuticle, he will escape detection. Vain delusion. Every person whose organ of smell is not wholly obliterated by snuff or catarrh, will single him out as quickly as a dog detects the exact locality of a weasel.

In the winter, one or two general baths each week will usually be sufficient to keep a person decently clean. But during the hot weeks of summer, a daily bath is indispensable. Two or three times a week, plenty of soap and water should be employed. On other days, a light sponge or towel bath will answer. A large quantity of water is not always absolutely necessary. A person can take a very refreshing and useful bath with a soft sponge and a pint of water. Such a bath can be taken anywhere without the slightest danger of soiling even the finest carpet. A simple air bath is better than none.

Cold bathing is not recommended. Robust persons may stand it very well, but it is injurious to invalids, and to any one if long continued. The best temperature for most persons is about blood heat.

Are not baths weakening? The weakening effect of a simple application of a little water to the surface of the body is not one-tenth as great as that from carrying about constantly a load of dirt upon the skin which not only prevents the elimination of impurities from the blood, but is actually absorbed into the system again. A bath is refreshing, soothing, and strengthening, if properly taken.

Tight-Laced Fissure of the Liver.—We once found in Bellvue Hospital, New York City, a woman who was suffering under a complication of maladies which evidently had their origin in the foolish practice of tight-lacing to which she had been addicted. On making an examination of the internal organs, we were amazed to find the liver presenting itself just above the hip bone, its normal position being entirely above the lower border of the ribs. Further examination revealed the fact that in about the middle of the organ there was a constriction, or fissure, nearly dividing it in two, which had been produced by habitual lacing. The function of the organ had been so greatly interfered with that it had failed to remove the biliary elements from the blood, and they had been largely deposited in the skin, making the latter anything but beautiful, although the woman was not advanced in years, and was naturally fair. Thousands of young ladies have cut their livers nearly in two in the same way. No wonder that they require rouge and French chalk to hide their tawny skins.

Thin Shoes.—Illy-clad feet are not infrequently the cause of very serious disease. A tight shoe prevents the proper circulation of the blood in the foot, causing it to become cold. If the shoe or boot is thin, the foot is still further chilled, and the blood which circulates with difficulty through it is sent back to the internal organs with a temperature much below that required for health. Exposure to cold causes the blood-vessels to contract so that less blood can circulate through them. Thus, one evil creates another. Thin soles, being insufficient protection against wet, allow the moisture of damp walks to reach the feet, making them wet as well as cold. When the extremities are chilled, the internal organs and the brain become congested, too great a quantity of blood being crowded into them. This is the chief origin of the headaches from which school girls suffer so much, but which are usually attributed to study.

Keep Warm.—Fashionable dress totally disregards every consideration but novelty and display. Fashion loads the shoulders and chests of ladies and girls with warm shawls, cloaks, and furs, surrounds the abdomen with ten to fourteen thicknesses of cloth, and imprisons the hands in an enormous muff, but leaves the limbs and ankles exposed to chilling blasts almost without protection, while they actually need more clothing than any other part of the body.

The whole body should be clad in soft flannel from neck to wrists and ankles nearly the year round. It is better to have the under-clothing for the upper part of the body and that for the limbs combined in one garment. If arranged in two garments, they should only meet, and not overlap, as this gives too much additional heat over the abdominal organs. A woman's limbs require as many thicknesses as a man's; and a garment which fits the limb closely will afford four times the protection given by a loose skirt. Thick shoes or boots with high tops, and heavy woolen stockings which are drawn up outside the under-garments clothing the limbs, complete the provision for warmth. Leggins should be worn in cold weather.

Squeezed to Death.—Not long ago a young lady went to bed without removing her corset, as she wished to grow small. When morning came, her friends found her a lifeless corpse. Thousands of young ladies are killing themselves in the same way. They may not die as suddenly, but they are dying as surely.

If any young lady who wears a corset could see the terrible havoc which it makes among her internal organs, she would be ready to desist from so foolish and harmful a practice. If the opportunity were afforded her, she would see her stomach squeezed out of shape and position so as to resemble much more a dog's than a human stomach. She would find her lungs compressed so that the blood could circulate with freedom through only a small portion, while the heart must struggle to its utmost to secure even a partial circulation. The large and small intestines she would find all jammed down into a heap in the lower part of the abdomen, where they do not belong, crowding upon the most delicate organs of her whole body, displacing and otherwise injuring them.

Any young woman who can deliberately commit all of these assaults against her physical frame while knowing the consequences, is guilty of a crime different from that of the suicide only in degree.

Night Air.—A general prejudice exists in the world against night air. In part it is justifiable; but much of it is unfounded. There is only one kind of air in the night, and that is night air. The air in the house is night air as much as that out of doors. All the air we breathe comes from the outside. If the windows and doors are shut, it crowds in through the cracks and chinks. It makes very little odds, then, whether we breathe night air in-doors, or out-of-doors, except that it is rather purer in the latter situation. In many localities night air is purer than day air.

Hygienic Agencies.—Nature has not provided agents by the use of which the penalty of transgression of her laws may be evaded; but there are certain natural agents, the proper employment of which will preserve health. If a person becomes diseased by neglecting to thus use these health-promoting agents, the only proper, and most efficient, way in which to recover from disease is to commence at once to do that which has been neglected. Thus it is that those agencies which are promotive of health and life become remedies for disease.

As might be supposed, from the foregoing, the

most potent remedies must be those agents which are the most essential to the maintenance of life and health. Among these, the following are the chief:—

Air, water, food, clothing, exercise, rest, cheerfulness, sunlight, and electricity.

Air.—Pure air is the first and the last desideratum of human life. Individual life begins with the first breath, and ends with the last act of respiration. A human being lives largely in proportion as he breathes. Frogs and lizards are sluggish because they breathe little. Birds are more vigorous in their movements because of the wondrous capacity and activity of their lungs. So with human beings. Need we suggest that those feeble-minded creatures who emulate each other in compression of the waist—thus curtailing their breathing power—are like frogs and lizards in their capacity for appreciating the "joy of living"? or that their organs of cerebration may be as small as their waists? Has a man consumption? Let him live in the open air; he cannot breathe enough. Thousands of patients die in hospitals for want of fresh air. God's oxygen is the best tonic known. Fill the sick-room with it; the patient's chances for recovery will be thereby increased fourfold. Its disinfectant and deodorizing properties are unsurpassed. All it requires is unrestrained action.

Water.—This limpid fluid constitutes three-fourths of the whole weight of the human body. The brain, the organ of thought, contains a still larger proportion. Its value as a curative agent is in direct ratio to its importance in the structure of the body. Water is valuable, 1. To dilute the blood, being the *only* drink ; 2. To cleanse the body from impurities within and without ; 3. As the most efficient means of applying heat and cold in the various forms of baths. Nothing relieves thirst like water. Nothing will regulate the temperature of a fever patient so effectually as water applied in the form of a cool pack. In relieving the coma of narcotic poisoning, apoplexy, sun-stroke, and lightning stroke, cold affusion is more potent than all other remedies combined. No salve, liniment, plaster, ointment, or medicated lotion is equal to pure soft water as a dressing for wounds. Water—hot, warm, tepid, cool, cold, or iced—is useful at the proper time.

Food.—" As a man eateth, so is he." A loaf of bread, eaten, digested, assimilated, becomes flesh. A pound of pork, treated in the same way, also becomes flesh. The first becomes pure, healthy flesh ; the second becomes gross, diseased flesh. Lord Byron appreciated this fact when he declared that he " felt himself grow savage " whenever he partook largely of animal food. If a man has filled himself with grossness, so that

his liver is clogged, his stomach and bowels torpid, all his vitals congested, and his life-current
sluggish, the best and only remedy is to "mend
his ways" at once and adopt the diet which nature indicates is best.   In this way thousands of
wretched dyspeptics and hypochondriacs have
sought and found their squandered health.   Try
it, reader.

Clothing.—The absurdities of fashionable
dress are too glaring to require exposure. All
admit the need of reform, but few have moral
courage to break Dame Fashion's shackles.   To
the pinioned, corseted, panniered, fettered, drag-
ged-down, tied-back, gasping, dying daughter of
Fashion, who would scarcely be conscious of living except for the aches, pains, nerves, neuralgias, stifled sighs, palpitations, and hysterics
which make up her wretched existence, what an
emancipation is offered in a dress which clothes
the body equably from head to toe! gives
perfect liberty of action to every muscle! allows room for a deep inspiration and a vigorous
heart-beat! removes from the hips those cum-
brous, dragging weights, and unties the lower
extremities!

Exercise.—Life is activity. Stagnation is
death.   This is true everywhere.   It is this
alone that makes the difference between the
sparkling brook, and the slimy pool; the bloom-
ing flower, and the withered shrub; the labor-

er's brawny arm, and the student's flaccid muscle. Few men die of excessive brain-work; many die from lack of muscle-work. Proper exercise is a powerful remedial agent.

Rest.—During sleep is the time when nature converts her work-shop into a repair-shop, mending broken nerve fibers, replenishing wasted muscles, repairing tissue cells, and renovating worn-out particles. When the body is wasted by disease, how much of this work there must be to be done! and how important that sleep be afforded as a prerequisite for its accomplishment!

Cheerfulness.—"Laugh and grow fat" is an old adage. *Laugh and get well* would be just as true. Indeed, the remedial power of a hearty laugh is sometimes greater than that of any drug in the *materia medica;* and its salutary effects have often saved the life of a failing patient. "A merry heart doeth good like a medicine" is good "Bible hygiene."

Sunlight.—Sunshine paints the skies, colors the leaves, and tints the flowers. Under its genial influence all nature thrives. It surpasses all other agents in restoring a natural color to the blanched and ghostlike faces of long-housed invalids. Sun-baths are powerful remedies for disease if rightly used.

Electricity.—This subtle agent, which flashes

in the thunder cloud, and quivers in a drop of
dew, is equally potent for good or evil. When
rightly used, its curative value is immense ; but
it has fallen, unfortunately, almost entirely into
the hands of quacks, who not only do much injury by injudicious applications, but bring disre-
pute upon it by claiming for it that which is
palpably absurd, as that it is the " nervous fluid,"
" vital force," " life force," etc.

FOOD AND DIET.

A MAN is made of what he eats. Good food
and drink make good blood ; and good blood is
manufactured into healthy brains and strong
bones and muscles. Poor food and improper
drinks make poor and foul blood, which, in turn,
is made into equally poor brains, bones, and muscles.

Those who pay no attention to the character
of their food, but hurry into their stomachs, in-
discriminately, food which is good, bad, and in-
different, are sooner or later admonished by disease and suffering that the way of the transgressor is hard, and that nature's laws are inexora-
ble. America is known abroad as a nation of
dyspeptics. This unfortunate condition is the
result of the universal disregard of dietetic rules
for which our countrymen are notorious. At-

tention to a few plain principles would save many thousands of lives annually. A large number of the most fatal acute diseases have their chief cause in errors of diet.

Poor Food.—Impoverished food is that which does not contain all the elements of which the body is built up in proper proportion. Perhaps the poorest article of food in common use in this country is fine-flour bread. The miller removes the very best and most nutritious portion of the wheat by the process of bolting; for the gluten which nourishes brain and muscle is deposited around the outside of the grain, just beneath the horny covering, or bran. In the center of the grain is found almost nothing but pure starch, which is so incapable of sustaining life that even a dog will starve to death in a short time if fed upon it exclusively.

Of such material nearly all American bread is made. Most other nations are wiser in this respect than we. The sturdy German eats his black bread made of the whole grain with a keen appetite, and it makes his muscles firm and his sinews strong in spite of the pernicious influence of his favorite lager beer.

Wheat-meal or graham bread is incomparably sweeter, richer, cheaper, and healthier than that made of the superfine, bolted, impoverished article.

Condiments.—Every day a hundred thousand

dyspeptics sigh and groan in consequence of condiments. Pepper, spice, salt, vinegar, mustard, and all kinds of fats belong to the list of dyspepsia-producing articles known as condiments. All the works on diet define a condiment as an article which adds nothing to the real nutritive value of food. It is simply something which is added to make food taste better. Whether the food does taste better or not does not depend upon the condiment, but upon the taste of the eater. If his taste is unperverted, he likes food best without condiments. If his taste is perverted, he may like almost any kind of unnatural combination. A Frenchman is as fond of assafœtida in his food as an American is of salt, or an East Indian of curry powder.

Condiments are innutritious and irritating. They induce a heated condition of the system which is very unfavorable to health. They clog the liver, imposing upon it a great addition to its rightful task. Worst of all, they irritate the digestive organs, impairing their tone and deranging their function. A little practice soon accustoms a person to the disuse of condiments, and he learns to relish his food better without than with them.

Facts about Salt.—It is a general supposition that salt is indispensable as an article of diet. Many people suppose that life cannot be sustained without it: nevertheless there are numerous facts which indicate that this popular supposition is erroneous. The following are a few of the many that might be presented :—

1. *Salt is a mineral.* It is a well-established fact that animal life cannot be sustained by the use of inorganic or mineral substances as food. Vegetables subsist upon inorganic matter, while animals require organized matter for their food.

2. *Salt is an irritant.* And when taken into the system it produces irritating effects. This is indicated by dryness of the throat, and acceleration of the pulse.

3. *When taken into the system it is not used,* being expelled, unchanged, by the liver, kidneys, skin, and other depurating organs.

4. *It is an antiseptic.* And when taken in any considerable quantities it greatly interferes with digestion.

5. *It is not necessary to support animal life,* as shown by the fact that its use is confined to a very small minority of the animal kingdom.

6. *It is not necessary to sustain human life,* as is conclusively shown by several facts: *a.* Scores of people who have been accustomed to its use have wholly discarded it, not only without detriment to their health, but with positive improvement. *b.* Millions of human beings in Central and Southern Africa, in South America, in some portions of North America, in Siberia, and in other parts of the world, subsist entirely without salt. *c.* This is not altogether because

salt cannot be obtained ; for in Southern Africa, where salt abounds, neither human beings nor lower animals make any use of it whatever.

We would not recommend that salt should be wholly discarded in all cases ; but there can be no doubt that many cases of diseases of the stomach and liver originate in the excessive use of salt. Persons suffering with torpid livers will find great benefit by abstaining almost wholly from its use, together with that of other condiments.

A gentleman who has just returned from a visit to England, states that many of the English stock-raisers who are noted for producing the finest cattle in the world, never think of feeding their cattle salt, as is so commonly practiced in this country.

Vegetable vs. Animal Food.—It is a mistaken opinion that flesh food is necessary to maintain human life. This is abundantly proven by numerous facts which are drawn from the anatomy of man and the lower animals, human and comparative physiology, and the experience of the human race from Adam's time to our own.

Flesh food is not necessary to sustain either mental and physical vigor, or animal heat. It contains no nutrient element not found in vegetables. In fact, eating flesh is only taking vegetables at second hand for all animals subsist upon vegetables.

On the other hand, the use of meat is unfavorable to longevity. Flesh food is stimulating. It contains venous blood, which is filled with such poisons as urea, uric acid, and cholesterine, with many others which would have been removed by the kidneys and liver of the animal had it lived. It is also liable to contain the products and germs of disease; for few animals are perfectly healthy when killed, and many are in a condition of gross disease, being only hindered from dying a natural death by the intervention of the butcher's knife.

Animal food will sustain life, it will nourish the body; but it is not the best food. Science shows that it is not the natural food of man, and history testifies that the bravest and noblest nations of antiquity subsisted for ages without it.

Thousands of people have investigated this subject during the last twenty years, have become convinced that animal food is inferior to vegetable food, and have renounced the use of the former with the most excellent results.

Persons quite advanced in years, or in feeble health, unless they have special morbid conditions which demand such a change, should not attempt to discard animal food altogether. In such cases, if any change in dietary is made, it should be very gradual, and should be made to occupy a considerable period of time. Much harm has been done by extremists in advising consumptives and other chronic invalids to abstain totally from the use of meat. When the system is in a debilitated condition it is not prepared to adapt itself to radical changes in diet unless there exists an imperative demand for it.

A Live Hog Examined.—Look at that object in a filthy mud-hole by the roadside. At first you distinguish nothing but a pile of black, slimy mud. The dirty mass moves! You think of a reptile, a turtle, some uncouth monster reveling in his Stygian filth. A grunt! The mystery is solved. The sound betrays a hog. You hasten by, avert your face, and sicken with disgust. Stop, friend, admire your savory ham, your souse, your tripe, your toothsome sausage, in its native element. A dainty beast, is n't he?

Gaze over into that sty, our pork-eating friend. Have you done so before? and would you prefer to be excused? Quite likely; but we will show you a dozen things you did not observe before. See that contented brute quietly reposing in the augmented filth of his own ordure! He seems to feel quite at home, does n't he? Look a little sharper and scrutinize his skin. Is it smooth and healthy? Not exactly so. So obscured is it with tetter, and scurf, and mange, that you almost expect to see the rotten mass drop off, as the grunting creature rubs it against any projecting corner which may furnish him a convenient scratching-place. As you glance around the pen, you observe that all such conveniences have been utilized until they are worn so smooth as to be almost inefficient.

Stir up the beast and make him show his gait. See how he rolls along, a mountain of fat. If he were human, he would be advised to chew tobacco for his obesity, and would be expected to drop off any day of heart disease. And so he *will* do, unless the butcher forestalls nature by a day or two. Indeed, only a few days ago a stout neighbor of his was quietly taking his breakfast from his trough, and grunting his infinite satisfaction, when, without a moment's warning, or a single premonitory symptom, his swinish heart ceased to beat, and he instantly expired without finishing his meal, much to the disappointment of the butcher who was anticipating the pleasure of quietly executing him a few hours later and serving him up to his pork-loving patrons. Suppose his death had been delayed a few hours, as is the case with the majority of hogs? or rather, suppose the butcher had got the start of nature a *little*, as he generally contrives to do?

But we have not half examined our hog yet. If you can possibly prevail upon yourself to sacrifice your taste, in the cause of science, pork-loving friend, just clamber over into the reeking sty and take a nearer view of the animal that is destined to delight the palates of some of your friends, perhaps your own. Make him straighten out his fore leg. Now observe closely. Do you see an open sore or issue a few inches above his foot, on the inner side? and do you say it is a mere accidental abrasion? Find the same on the other leg; it is a wise and wonderful provision of nature. But what are they? Grasp the leg high up, and press downward. Now you see, as a mass of corruption pours out. That opening is the outlet of a sewer. Yes, a scrofulous sewer; and hence the offensive, scrofulous matter which discharges from it. Should you fill a syringe with mercury, or some colored injecting-fluid, and drive the contents into this same opening, you would be able to trace all through the body of the animal little pipes communicating with it.

What must be the condition of the body of an animal so foul as to require a regular system of drainage to convey away its teeming filth? Sometimes the outlets get closed by the accumulation of external filth. Then the scrofulous, ichorous stream ceases to flow, and the animal quickly sickens and dies unless the owner speedily cleanses the parts, and so opens anew the feculent fountain, and allows the festering poison to escape.

What dainty morsels those same feet and legs make! What a delicate flavor they have, as every epicure asserts! Do you suppose the corruption with which they are saturated has any influence upon their taste and healthfulness?

The hog is a scavenger by nature. His organization indicates it, for he has a regular system of sewers running all through his body and discharging on the inside of his fore legs, the express object of which is to convey away the filth with which his body teems.

The process of fattening hogs is one of disease. A fat hog is one which is grossly diseased. That this is the case is shown by the condition of the liver. The livers of all fat hogs are masses of disease. Every butcher will tell you that he finds not more than one liver in twenty among fat hogs which is not crowded with abscesses.

Tape-Worm.—This loathsome creature, which sometimes gets into a human stomach and intestines, and grows there to the enormous length of several rods, is communicated to man by eating pork. The occurrence of tape-worm is becoming much more frequent in this country than formerly, owing to the free use of pork.

Trichinæ.—Still more to be dreaded by pork-eaters are the terrible trichinæ, which are also communicated by the eating of pork. Each worm is so small that several hundred thousand of them may occupy a single cubic inch of pork. When taken into the body, a single worm produces ten young, which at once commence boring into the body in every direction, lodging at last in the muscles. The pain and general disturbance of the system is so great that few constitutions can survive the terrible ordeal. If life is not destroyed at once, the individual lingers along, a sufferer for life, his body filled with disgusting worms for which there is no remedy. No cure for the disease has been discovered. About one hog in every ten is affected by the disease. No more than one in ten of the deaths from this cause are attributed to it, as the disease may appear like many others, resembling cholera, dysentery, typhoid fever, cerebro-spinal meningitis, and rheumatism. No pork is safe.

Poisonous Water.—Whole communities have been stricken with disease at once by what seemed a very mysterious cause. Investigation traced the origin to the water supply. Further investigation proved that the original source was some sewer or privy which communicated with the water supply. This is known to be one of the greatest causes of typhoid fever.

The water of wells is often rendered poisonous by receiving the drainage of barnyards and vaults. Sometimes matter of this character will be conducted many feet under ground in a pervious soil, by percolation.

Water from a barnyard well or cistern should never be used. No vault or cess-pool should be within fifty feet of a well.

Milk from Stabled Cows.—Milk is not the best food, because it contains the impurities of the blood of the animal from which it is taken. If the animal's blood be pure, the milk is proportionately good; if it is impure, the milk must be likewise affected.

When cows are confined in a close stall, they breathe over and over the same foul air, which is always loaded with filthy vapors from their own excreta. These vapors enter the blood and poison every tissue and every secretion. The inhaled impurities make their appearance in the milk also, which thus becomes a means of excretion. If it is eaten, the filthy impurities of the stable are taken with it.

A writer of note truly says that "fully one-

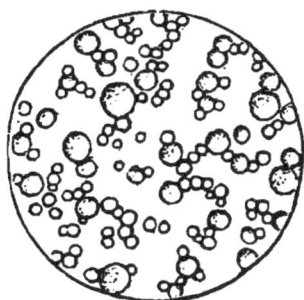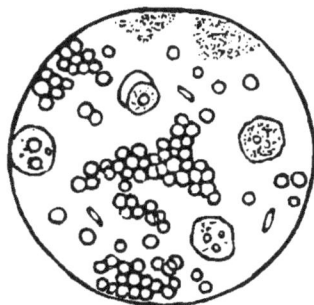

Fig. 1.　　　　　*Fig. 2.*

half of the deaths among the young are directly traceable to poisonous milk;" and yet thousands of people, especially in our large cities, are daily exposing themselves and their children to the possibility of fatal poisoning.

The taste is not always a reliable means for testing the quality of the milk, neither can the poisonous elements be detected by the closest scrutiny of the chemist; but the microscope reveals the presence of disease, although it may escape all other means of detection.

Fig. 1 is an accurate illustration of the appearance of pure milk when examined by means of a good microscope. It will be seen that it contains nothing but rounded globules of various sizes, which are the so-called butter cells of milk.

Fig. 2 is an exact representation of the appearance of diseased milk under the microscope. This specimen was taken from a cow that was fed upon swill and confined in a filthy stable. The difference between these two specimens will be readily observed. In Fig. 2, in addition to the rounded globules which are alone found in Fig. 1, we have great numbers of minute organisms which are indicative of disease. Milk of this kind cannot be habitually used without producing serious disturbances in the system.

Catching Consumption.—French experimenters have ascertained that cows are very liable to consumption, and that the tubercle of this disease may be communicated by eating either the flesh or the milk of affected animals. This will account in part for some of the cases of "quick consumption;" for it is observed that when the disease is communicated in this way its progress is much more rapid than under other circumstances. When milk is used, the greatest care should be taken to obtain it from healthy animals.

Poisonous Sirups.—For a number of years the people of this country have been abused by the manufacture and sale of villainous compounds which were labeled with such enticing names as, "golden drip," "silver drip," and similar phrases. These so-called sirups, instead of being made from sugar or the sugar-cane, are manufactured by chemical processes, being made from starch, cotton rags, saw-dust, and similar materials.

It has long been known to chemists that a sweet substance, known as grape-sugar, could be produced by boiling starch for a long time with sulphuric acid. Saw-dust, cotton, and woody fiber in any other form, furnish the same product when treated in a similar manner. Unscrupulous knaves have taken advantage of this scientific fact to impose upon the people a spurious kind of sirup. These unrighteous practices have become so extensive that it is next to impossible to find a specimen of sirup that is wholly free from contamination.

The effects of using this chemical preparation are very serious. It contains sulphuric acid, or oil of vitriol, iron, and various other unwholesome constituents. When freely used, it produces irritation of the stomach, and it has, no doubt, been the cause of numberless cases of chronic dyspepsia. In one instance which occurred under our observation, more than a dozen people suffered at once with slight symptoms of poisoning, the consequence of eating candy made of this wretched stuff. It was observed that the teeth and tongues of those who ate of the candy were made very black; and without doubt the blackened teeth were permanently and seriously damaged.

It is important to know how to distinguish these adulterated and poisonous sirups from those which are pure. A convenient method, which is sufficiently accurate for practical purposes, is to add a teaspoonful of the suspected sirup to a half cup of strong tea. If the solution becomes black, like ink, the sirup is unfit for use and contains poisonous elements. It should certainly be discarded. If every family would adopt the plan of testing sirup before buying, and refuse to purchase that which would not stand the test, the market for the vile compound would soon be destroyed, and its manufacture would necessarily cease.

Tea and Coffee.—One of the most common causes of dyspepsia, "liver complaint," and nervousness, is the use of tea and coffee. The injury resulting from the use of the beverages is attributable to several evils.

1. The active principle of both tea and coffee is theine, or caffeine, a narcotic poison, which is fatal in other than small doses. Although not fatal in small doses, it produces, nevertheless, a decidedly injurious effect. The full injury is not seen at once, neither does it appear in a few months; but the integrity of the digestive and nervous systems is steadily, though slowly, un-

dermined. Chocolate and cocoa occasion precisely the same effects, though they are less powerful.

2. The tannin contained in an infusion of tea or coffee disturbs digestion by rendering inert the gastric juice, one of the most essential digestive agents.

3. The use of hot liquid of any kind at meals is very damaging to the stomach. The organ is not only over-stimulated by the abnormal heat, but its function is impaired by excess of fluid. The gastric juice is diluted so much as to be rendered incapable of performing its function, and the stomach is wearied with the task of absorbing the superabundant fluid. Meanwhile, the food undergoes fermentative changes, and becomes unfit for nourishing the body.

Hundreds have found a cure for dyspepsia, sick-headache, nervousness and wakefulness at night, in discarding tea and coffee with all their substitutes.

Hard Water.—Water containing lime and other mineral matters is productive of several very painful diseases. Avoid its use. Soft water can always be obtained at certain times, and preserved in cisterns. Such water is only fit for use after filtering. (See directions for making a filter.) Boiling hard water removes a portion of the lime. Filtration does not purify it.

It is a mistaken notion that hard water is necessary for the maintenance of health. Nothing could be more absurd. The softest, purest water is the best for all the purposes for which water is needed in the human body.

Iced Water.—Copious draughts of iced water are very injurious. In the summer time especially, iced water is harmful on account of the sudden cooling of the internal organs which it induces. If drank at all, it should only be in small sips and very slowly.

The injudicious use of iced water in summer is a most common cause of dysentery and other bowel troubles. It also frequently produces a weakened condition of the digestive organs which results in dyspepsia.

Iced cream, iced tea, and iced milk, together with all other varieties of ices, should be avoided by those who have any anxiety to preserve the health of their digestive organs.

Eating Between Meals.—The stomach requires rest as well as the brain or the muscles. If food is eaten at other times than at meals, it is kept constantly at work. From three to six hours are required to digest most articles of food; hence, if food is taken again within five or six hours after eating, the stomach is kept incessantly employed, and becomes exhausted. When the next meal is taken, it is unprepared to receive it, and indigestion with its myriad train of ills results. Late suppers are suicidal. Never eat within five hours of retiring.

Hasty Eating.—Americans are proverbial for hasty eating. The student swallows his food unmasticated, and hastens back to his books. The merchant bolts his meal to save time for business. The glutton eats as fast as ever he can to keep pace with his neighbors and get his full share.

It is not enough to fill the stomach with food. Digestion begins in the mouth ; and unless the mouth does its share of the work, the stomach is required to do a double portion. When the food is sent down into the stomach in lumps, the abused organ does its best to digest it, but fails, because it has no means for grinding food. The mill is in the mouth, and mastication, if done at all, must be done there. The gastric juice cannot act upon solid food, and allows it to go undigested. Fermentation ensues, and dyspepsia, dysentery, cholera morbus, and a dozen other diseases result.

Eight ounces of food, well masticated, will afford as much nourishment to the body as a pound hastily bolted.

Alcoholic Drinks.—No well man can habitually use wine, beer, brandy, or any other alcoholic drink, without becoming diseased. It is good for nothing as a food, and is useful as a medicine only when used with great discretion. Old people do not require it any more than young persons. Indeed, it is far more dangerous for old than young, because it renders them liable to apoplexy.

Moderate drinking is a skillful trick of the old serpent to lead men to drunkards' graves.

Any quantity of alcohol intoxicates. Intoxication is poisoning. A little alcohol intoxicates a little; a larger quantity intoxicates a good deal. The moderate drinker, no matter how small his libations, only differs from the gutter toper in degree.

The following "Facts about Alcohol" are well worth the consideration of those who need to be warned of the consequences of becoming addicted to its use :—

Facts about Alcohol.—1. Alcohol is a poison. When pure, it will produce death as certainly and almost as quickly as prussic acid.

2. Alcohol is a product of fermentation, or decay. The Creator never made it. No plant produces it. No bubbling spring affords it.

3. Alcohol is an irritant. It will blister the skin, and produce inflammation of the stomach.

4. Alcohol is a narcotic. It paralyzes the nerves, and benumbs the sensibilities.

5. Alcohol destroys the blood. It dissolves the blood corpuscles, and thus impoverishes the vital fluid.

6. Alcohol causes heart disease, by changing the heart tissue for fat.

7. Alcohol causes apoplexy. It weakens the

blood-vessels, and causes congestion of the brain. Alcohol weakens the muscles. It has been proven by experiment that a man can lift less after taking a glass of whisky than before.

8. Alcohol wastes vital force.

9. Alcohol causes consumption.

10. Alcohol lessens bodily heat. Travelers in the Arctic regions are obliged to be teetotalers.

11. Alcohol causes paralysis of the brain. A man who is dead drunk is temporarily paralyzed.

12. Alcohol hardens the brain.

13. Alcohol produces congestion of every organ of the body.

14. Alcohol hardens the liver, and renders it useless.

15. Alcohol produces its worst effects when taken in small doses.

16. Alcohol produces all kinds of nervous disorders.

17. Alcohol occasions cancer, ulcer, dyspepsia, and other diseases of the stomach.

18. Alcohol is the cause of more than two-thirds of the cases of disease found in the hospitals in large cities.

19. Alcohol is one of the greatest causes of pauperism.

20. Alcohol is one of the most active causes of crime. In Scotland it increased the frequency of crime 400 per cent.

21. Alcohol is a great cause of insanity.

22. Alcohol shortens life 500 per cent., accord-

ing to the statistics of life insurance companies.

23. Alcohol annually kills one hundred thousand American citizens.

24. Alcohol costs more than bread.

25. Alcohol serves no useful purpose in the human system, and does inestimable harm.

Effect of Diet on the Liver.—Almost every other man we meet is complaining about his liver. One has a "torpid" liver; another has "congestion" of the liver; another has a pain in his side, which he is confident is due to disturbance of his liver. Complaints are loud and general against the liver, but no one thinks of entering a complaint against the diet, which is the real source of difficulty. Careful investigation and examination of the liver, after death, have proven the deleterious effect which certain articles of food have upon the liver.

The drunkard's liver becomes hardened by the alcohol which he imbibes. The liquid poison has the same damaging effect upon his brain.

The livers of people who use a great deal of fat—fat meat, butter, lard, rich cakes, pies, etc.—become infiltrated with fat. They undergo a process called fatty degeneration, in which there is an actual change of the tissue to fat. This change is favored by sedentary habits. The liver of the domestic cat is almost always fatty.

The natives of the East Indies, as well as of Central and Southern Africa, together with Mexico and other warm climates, make great use of pepper, mustard, turmeric, and other irritating spices. The result of this practice is not only derangement of the stomach, but the production of induration of the liver, a disease which was formerly attributed to the climate of those regions, on account of its prevalence, but is now well known to be the result of the use of the deleterious articles named. Lovers of pepper and mustard should look out for their livers.

It has been observed that cattle that have been overfed, or fed on warm slops, have badly diseased livers. The organ is found enlarged, in some cases very greatly, and its surface is covered with red spots and ragged, ulcerated patches, indicating the presence of disease of so extensive a character as to render the organ almost wholly useless.

The same causes which produce these grave effects in savage and semi-civilized human beings, and in lower animals, will produce the same results in civilized beings. Pepper and mustard are no better for a New York City gormand than for a Hottentot or a Mexican Indian. Slop food—highly seasoned soups, gravies, and " rich " sauces—will work for human livers the same mischievous results that follow its use by lower animals.

Two Meals a Day.—According to Hippocrates, a very noted Grecian physician who lived a few centuries before Christ, the Grecians of that age ate but one meal a day. He advised, however, that two meals should be eaten, as by so doing there would be less liability to overeating. Thus it is evident that the " two-meal system," as the custom of eating two meals a day is called, is not by any means a modern innovation, but has the sanction of antiquity. It is also a fact worthy of mention in this connection, that the ancient Grecians were among the most hardy, energetic, and courageous, as well as learned, of all the nations of whom we have any historical record. Their feats of physical prowess astonish the world ; and their rank as thinkers was in no way inferior to that of any other people who have ever lived. The advantages of two meals instead of three are very numerous ; and there are no substantial objections to the practice in any but a few exceptional cases. This is a favorable season of the year in which to begin the omission of the third meal. The change may be made at once, or gradually. Perhaps the latter plan is the better one for most persons. If breakfast is taken at $7\frac{1}{2}$ or 8 A. M., and dinner at 2 P. M., the supper will not be missed, or very little at most, especially if the individual retires early.

Of course there are cases in which three meals a day, if the supper be light and early, are preferable to a less number, and for such two meals are not recommended.

It would have been infinitely better for human

stomachs if the ancient custom of eating but twice in a day had been maintained. There are a great many other directions as well in which modern practices are no improvement over ancient ones, and which call for reform by a return to the customs of our predecessors.

Tender Meat.—Those who use animal food are always desirous of obtaining "tender" meat. In order to satisfy the demand for such food, the butcher and the producer resort to all sorts of devices. The former keeps the flesh of slaughtered animals after they are killed until decay has begun, in order that the natural firmness and elasticity of the tissues may be overcome by processes of decomposition. The latter treats his animals in such a manner previous to their death that their tissues become softened and disintegrated by disease. There are several means employed to effect this; chief among them are confinement and overfeeding. An exchange gives the following translation of a description of how young pigeons are fattened in Germany, as given in the North German *Allgemeine Zeitung* :—

"In order to fatten young pigeons quickly, put them, on the twentieth day, or when they commence to get feathers, into a basket with a soft layer of moss or hay on the bottom, in a place which freely admits the air, but excludes the light. Feed the birds three times daily, at intervals of five hours each, with cooked maize,

opening their beaks and making them swallow
successively thirty to forty grains each. The
maize should be warm, but not hot. By con-
tinuing this treatment ten or twelve days, the
birds will become most tender and delicate."

Such meat would doubtless be "tender"
enough to suit the most fastidious epicure. In
this respect the plan suggested would certainly
be perfectly successful; but great care would be
necessary lest nature should succumb and actual
dissolution of the poor birds occur before their
heads were chopped off. Mr. Bergh would arrest
the perpetrators of such cruelty.

Lager Beer as Food.—After such repeated
refutations of the idea, it is strange that people
should still cling to the notion that lager beer is
nourishing. If a man has lost his appetite, and
seems to be failing in strength, or losing weight,
his next-door neighbor advises him to drink
daily a few glasses of lager beer. If a nursing
mother has insufficient food for her infant, wise
old ladies prescribe lager beer or ale.

Although it is being constantly reiterated in
the ears of the people that alcohol is not food,
and that beer and ale are only dirty mixtures of
alcohol and water, still they refuse to believe
that these pernicious beverages cannot, in some
way, impart nourishment and strength. Perhaps
the testimony of one of the greatest of European
savants will correct the opinions of a few.

Said Prof. Baron Liebig, a German chemist of great renown, " We can prove with mathematical certainty that as much flour or meal as would lie on the point of a table-knife is more nutritious than five measures [ten quarts] of the best Bavarian beer." Powerful nutriment, indeed !

A Barbarous Practice. — The practice of smoking, which has now become so universal among a large proportion of our male population, has a curious and interesting history—curious, on account of the novel origin of the habit, and interesting, from the insight which it gives into the depravity of human nature.

For a long time, the origin of smoking was obscure ; but history has come to the rescue, and now we learn that " in 1492, as Columbus lay with his ships beside the island of Cuba, he sent two men to search the land and report what they might see. On their return, among other things, they said they saw the naked savages twist large leaves together, and smoke like devils." Since that time, a large share of the men and boys of civilized nations have been following the filthy example of those naked savages.

It was not, however, without meeting with vigorous opposition that tobacco obtained despotic tyranny over human beings. In Russia, the use of tobacco was prohibited under the penalty of the bastinado for the first offense, loss of the nose for the second, and deprivation of life for the third.

In Italy, the pope fulminated a bull against the filthy weed, and excommunicated all who used it in church.

In Switzerland, tobacco-users were treated as criminals.

The Shah of Persia made tobacco-using a capital crime, and many of its devotees were executed.

In Constantinople, a Turk was led through the streets with his nose slit and transfixed by a pipe-stem, as a warning to smokers.

King James I., of England, expressed his opposition to the weed in a powerful "Counterblaste," which stigmatized the drug in most decided terms.

Even in this country, the native home of tobacco, at a somewhat later period its use was interdicted to all who had not previously acquired the habit, unless prescribed by a physician as a medicine.

But the devotees of this fascinating drug steadily increased in spite of all opposition, until tobacco-using has become an almost universal vice; in which fact we see a striking illustration of the readiness of human nature to seize upon anything which promises gratification of the senses, no matter how filthy, how disgusting, how pernicious, or how fatal in its ultimate consequences.

Diet and Mental Labor.—Isaac Newton performed his most severe intellectual labor while subsisting upon a diet of bread and water. Pythagoras, one of the most acute philosophers of antiquity, was a rigid vegetarian, and educated his followers in the same regimen.

Cheerfulness at Meals.—The benefit derived from food taken, depends very much upon the condition of the body while eating. If taken in a moody, cross, or despairing condition of mind, digestion is slower and much less perfect than when taken with a cheerful disposition. The very rapid and silent eating too common among Americans, should be avoided, and some topic of interest introduced at meals, in which all may participate; and if a hearty laugh is occasionally indulged in, it will be all the better.

Spices.—The almost universal fondness for spices is a curious illustration of the readiness with which the simplicity of the natural taste may become depraved. Pepper was used before B. C. 400. Pliny speaks of its use in his day, and expresses his astonishment that men should esteem it so highly when it has not a sweet taste, nor attractive appearance, nor any other desirable quality. We can heartily sympathize with Pliny in his astonishment.

Nutmegs and mace are quite extensively used as spices in this country and in Europe; but neither one is ever used as a condiment in the country from which they were first brought, the Isles of Banda.

Simple Remedies

FOR COMMON DISEASES.

A LARGE share of the cases of illness which are constantly occurring in nearly every family are of such a character that they can be treated by any intelligent mother quite as well as, or even better than, by the doctor. Again, the necessary trouble of going for a physician for every trifling ailment, besides the useless expense in fees which it occasions, is a weighty consideration. Important cases demand medical advice; but every parent ought to be sufficiently well informed to be able to attend promptly and efficiently to the great majority of the ailments to which all families are liable.

If children are properly clothed and fed, allowed plenty of exercise, fresh air, and sleep, they will be seldom ill. The same is equally true of grown people. Accidents, exposures, and indiscretions will occur, however, resulting in various ailments. If the simple directions given for treating some of the more common diseases are carefully followed, much trouble, expense, and suffering may be avoided. Few drugs are recommended for internal use, because the cases in which they are really needed are such as require the personal attention of a physician.

Colds.—Tommy, or Mary, or baby, or some
other one of the children, or the family, has
"caught a hard cold;" what shall we do? Do
nothing, and let it wear off?

No; perhaps he will get well, may be his cold
will become something worse.

Shall we give him ginger tea, red pepper,
brandy sling, onion sirup, honey and lard, fat
pork, castor-oil, licorice, hoarhound, molasses
candy, boneset, catnip, mullen tea, or pennyroyal?
or shall we apply a mustard plaster to his chest,
a blister to the bottom of each foot, and fat pork
with salt and pepper to his throat?

Do no such thing. Such trash put into his
stomach, with such irritating applications outside, would make a well person sick. Now do
this :—

In the first place, prevent the cold, if possible,
by beginning in season. Perhaps the feet have
been wet, and are damp and cold. Pull off the
shoes or boots and stockings, and put the feet
into a pail of water as hot as can well be borne,
after first wetting the head with cool water.
After fifteen minutes' soaking, pour a little cold
water into the pail. Allow the feet to remain
two or three minutes longer, then take out, wipe
dry every part, between the toes and around the
ankles, and then rub them until they glow with
warmth. Put on dry, warm stockings, and send
the patient to bed for an hour, or all night if it
is evening. Instead of waking up in the morn-

ing with a headache, a sore throat, and a voice like a cracked fiddle, he will be quite well.

If a person has really got a cold, and is sneezing, and wheezing, and coughing, and expectorating, more thorough measures must be taken.

1. Eat little or nothing for a day or two. The popular adage, "Stuff a cold and starve a fever," is without foundation. A cold is a fever—a *heat*, really, rather than a *cold*, if temperature be considered.

2. Rest. Sleep all that is possible. No time is lost in such a course. Timely rest may save serious illness.

3. Take some kind of hot bath, which will start the perspiration freely. Long sweating is debilitating, only start the action of the skin. The foot-bath combined with the sitz-bath, the wet-sheet pack, the vapor-bath, and the hot-air bath are alike suitable. These are severally described in this work. After the bath, go to bed.

Drink freely of water, the purer the better.

A day or two of such treatment will usually "break" the hardest cold, saving the patient several weeks of pain and annoyance, if not from chronic disease. Try it. The trouble is less than you think, and the results are splendid.

Frequent bathing in tepid water makes a person less liable to colds.

Sore Throat.—There are many remedies for sore throat, some of which are harmless, being

simply worthless—like goose-oil applied exter-
nally—while others are quite injurious. The
remedy used by the Germans—and many sensi-
ble Americans—is the best. If it is a case of
simple sore throat, make, alternately, hot and
cold applications, according to directions given
elsewhere. If there is fever, cool the skin with
sponge-baths. Keep the feet warm. If there
are symptoms of diphtheria, apply ice in a bag to
the outside of the neck, and give the patient little pieces of ice to swallow. Lemon juice applied
to the pharynx with a swab is sometimes a good
remedy.

Hoarseness.—All the sirups, expectorants,
cough mixtures, anodynes, and inhalations ever
invented or advertised will not cure hoarseness.
They may sometimes destroy the sensibility of
the nerves of the diseased part, and so relieve
the cough, but they cannot remove the disease.
Honey, loaf-sugar, and all such articles are very
deceptive remedies. Cough lozenges and candy,
troches, etc., are equally useless. They do not
come in contact with the diseased surfaces, as
many suppose. They pass directly down into
the stomach, where they occasion much disturbance, disordering digestion, and so producing a
disease really worse than the one they were intended to cure.

If the disease has not become chronic, it may
usually be relieved by bathing the throat and

neck in cool water, applying heat and cold alternately, and wearing a wet bandage around the neck nights. If the difficulty is of long standing a physician's care is needed.

Headache.—Pain in the head is caused by either too much or too little blood. If the pulse is high and the head hot, while the feet are cold, apply cold to the head and put the feet in a hot bath. A sitz-bath and foot-bath combined will be necessary in severe cases. If the cold application does not give speedy relief, apply hot fomentations for a half hour, unless relief is sooner obtained, renewing the application every four or five minutes. Apply a tepid compress last.

Sometimes headache is caused by undigested food in the stomach. In such cases a warm-water emetic is needed. If accompanied by cramp in the stomach, apply fomentations over that organ also. Sick headache nearly always requires hot applications.

Burns and Scalds.—Apply at once light cloths dipped in cool or tepid water, or immerse the part in water. When the pain is somewhat relieved, apply pure lard or sweet-oil. One of the best preparations is sweet-oil to which carbolic acid has been added in proportion of one part to twenty. It may be applied by means of a saturated cotton or linen cloth laid over the part. If the burn has not destroyed much of the skin, prompt relief will usually be obtained by covering the part with the white of egg applied with a soft brush. Apply a second coat when the first dries. Deep burns should be poulticed after the pain has been somewhat relieved by the application of cool wet cloths, as they will be attended with sloughing and discharge of pus.

Alum-water and carron-oil (a mixture of lime-water and linseed-oil in equal parts) are favorite remedies with some. A saturated solution of bicarbonate of soda, applied by means of a thin compress, is recommended as a most excellent remedy.

Chilblains.—This troublesome affection, though seemingly insignificant, often makes existence almost a burden by its constant irritation. It is easily cured, but not by the application of any sort of salve, ointment, liniment, or quack nostrum, no matter how highly recommended.

Just before retiring, prepare two vessels for a foot-bath. Place in one, water as hot as can be borne, and in the other, very cold water. Place the feet first in the hot water for two minutes, then in the cold water for the same time. Alternate thus four or five times, merely dipping the feet in the cold water the last time, and then wiping them dry. Repeat this treatment every night until the cure is effected. Improvement will usually begin at once.

Wear thin cotton stockings inside the woolen ones, and avoid exposing the feet to severe cold until they are well. A general bath twice a week is necessary. (See article on freezing, for prevention of chilblain.)

Pain.—Acute pain is usually due either to inflammation or neuralgia. Hot applications are nearly always the most grateful and the most successful of any local remedy. Plasters, liniment, and leeches are seldom if ever useful. Blisters are wholly unnecessary, and are always harmful. The most judicious physicians have wholly discarded them. Sometimes cold applications are the most grateful and efficient. The patient's feelings will determine which is to be employed. The hot foot-bath, or the foot-bath and sitz-bath combined, is sometimes necessary in addition to local measures.

Face-ache.—Pain in the face is generally of a neuralgiac character. Frequently it originates in a diseased tooth. Make hot applications in any of the several ways described in the article on "Hot Applications." Cold applications are occasionally best. The foot-bath, sitz-bath, and abstinence from food are useful auxiliaries of treatment. When due to constitutional causes, as the use of tea, coffee, tobacco, or liquor, or to an impoverished condition of the blood and general derangement of the nerves, the disease is very obstinate and requires constitutional treatment.

Toothache.—This painful affection is often closely connected with face-ache. It may be due to a decayed or ulcerated tooth, or to disease of the dental nerve. Apply the same remedies as directed for face-ache. In addition, apply half of a steamed fig (hot) to the diseased tooth. A bit of cotton saturated with laudanum or creosote, and crowded into the cavity of a carious tooth, will often give speedy relief. The only proper and permanent remedy when the tooth is decayed, is to have it filled or extracted. It should be filled, if possible.

Earache.—Hot applications, or the prolonged hot douche, applied with the fountain syringe, will often give relief. A hot poultice, continually applied, and frequently changed, is a good remedy. Half a boiled or roasted onion, bound upon the ear, will sometimes give relief. No remedy is infallible. The hot foot-bath and sitz-bath are excellent remedies. If an abscess is forming in the outer ear, the pain will continue until it opens, or is lanced. A few drops of laudanum placed in the ear give relief in some cases, and can do no harm. A still better application is obtained by evaporating the alcohol from a teaspoonful of laudanum and mixing the residue with half a teaspoonful of sweet-oil or glycerine. Incline the head and pour a few drops of this into the ear. Such applications give relief only by deadening the sensibility of the nerves, and not by removing the cause of the difficulty. Hence, they should be employed, if at all, only in connection with other remedies.

Rheumatism.—Inflammatory rheumatism requires the attendance of an experienced person. The wandering pains from which many people suffer, which are commonly called rheumatism, can be relieved by proper attention.

1. Avoid the use of irritating condiments, tea, coffee, tobacco, and alcoholic liquors, including wine, beer, etc. Avoid, also, gross food, and the use of food or drink containing saline matters. Be temperate in all things.

2. Dress warmly and uniformly. Silk or buckskin under-suits, worn next the cotton under-clothing, give great relief to many. Wear flannel the whole year.

3. Apply heat to the painful parts as in neuralgia. The hot-air and vapor-baths are good. Keep the skin clean. Exercise freely.

Colic.—The usual causes are indigestion and constipation. Administer a copious enema to secure a free passage from the bowels. Apply dry, hot cloths or hot fomentations over the abdomen. Percuss and knead the abdomen gently, to promote action of the bowels. Hot drinks do very little good, and usually as little harm. For an infant, fold a thick woolen blanket, wet one end in as hot water as can be borne, wring it so that it will not drip, and apply the wet end over the abdomen of the child, wrapping the remainder around its body. It is often surprising to mark the almost instantaneous relief which follows. The applications must be *hot*, not simply warm, and must be renewed every five or ten minutes until relief is obtained.

Nearly all abdominal pains may be relieved in the same way.

Convulsions.—The convulsions of children—commonly called spasms, or fits—are usually due either to worms or indigestion, unless they occur in the course of some acute disease. Place the child at once in a hot bath, disturbing it as little as possible. It will usually recover in a few minutes. When sufficiently recovered, administer an enema to free the bowels, and keep it perfectly quiet. Some advise the cold bath, and practice it with good success. The patient should be rubbed vigorously during the cold bath.

Epileptic convulsions require more than simple domestic treatment. The most that can be done for the patient during the fit is to prevent him from injuring himself or others. The lips and tongue are often severely bitten by the spasmodic action of the muscles of the jaws closing the teeth together upon them. This may be prevented by placing a piece of soft wood or other material between the teeth at the beginning of the fit. As the patient usually sleeps some time after the fit, the brief interval of consciousness which immediately follows it should be occupied in getting him into a comfortable position.

Hysterics.—This peculiar disease is most common in women, though sometimes observed in men. It is a real disease, and should be treated as such. The symptoms are almost as various as the cases. It may simulate any disease. Place the patient upon a sofa, beside which a large vessel is placed. Hold the head of the patient over the vessel, and pour cold water upon it from a pitcher held a few feet above. Apply at the same time cold to the chest and spine, and hot bricks or bottles to the feet. This treatment may be continued for an hour or two without injury if the patient does not recover sooner. Speedy relief is usually secured. If the patient becomes quite chilly, apply warm cloths to the chest and shoulders.

Apoplexy.—If a person falls suddenly and is found with a full pulse, throbbing temples, flushed face, and breathing hard, he has apoplexy. Loosen every constriction about the throat at once, elevate the head, secure fresh air, bare the chest, and pour cold water upon the head. See that the extremities are warm. Call a physician as soon as possible. Do not bleed, nor give brandy, ammonia, nor any other stimulant. Apoplectic convulsions are quite rare. They generally occur in sedentary people of full habit, in advanced life.

Fainting.—When a person faints, the heart nearly ceases its action, the action of the lungs is nearly or quite suspended, the face becomes pale, and partial or complete unconsciousness ensues. If the person has fallen, do not elevate the head, but be careful to keep it as low as, or lower than, the rest of the body. If the patient is sitting in a chair, step behind him, grasp the chair at the sides, and carefully tip it back until the head touches the floor. This alone will suffice in many cases. If the patient does not immediately revive, loosen the clothing about the neck, chest, and abdomen ; sprinkle cold water in the face ; slap the surface of the body with the hand or a slipper ; apply an ammonia bottle, camphor, or any other pungent odor to the nostrils ; secure abundant cool, fresh air, and use artificial respiration. If the patient can swallow, give very hot or very cold drinks.

A person who is subject to syncope should lie down at once when he first feels faint.

Croup.—If the child can speak aloud, the disease is of the spasmodic variety, and he will probably recover with a little attention ; but if he can only whisper, and the disease has come on somewhat gradually, it is a much more serious variety—true croup—and a physician should be called at once.

Apply, alternately, hot and cold cloths to the throat and neck for a half hour, then apply cold continuously for half an hour, then foment again. Give a hot bath, and keep the limbs and extremities warm. Give no emetics, expectorants, stimulants, nor anodynes ; all are harmful. Goose-oil on the outside does no more good than ipecac inside. Give the child an abundance of fresh air. If the case is one of true croup, the inhalation of steam is one of the best remedies.

Measles.—Ordinary cases require little more than care and good nursing. The comfort of the patient is greatly increased by frequent tepid sponge-baths or packs. If the eruption does not appear promptly, or is repelled, put the patient into a hot pack, with a woolen sheet, for thirty minutes. Keep the head constantly wet with cool water, and bathe the face every few minutes when there is considerable fever. If the throat is sore, give treatment for sore throat as already described. Give the patient abundance of fresh air, but do not expose him to draughts. The diet should be as simple as possible, and very light. Slings, teas, sirups, and other medicinal agents are not required in this disease.

Scarlet Fever.—This disease may be treated essentially in the same manner as measles. The sponge-bath should be administered several times a day. Keep the bowels free by enemas.

Fevers.—Simple fevers may be treated in accordance with the directions for measles and scarlet fever. If complications occur, as pleurisy, lung fever, or other affections, a physician should be consulted.

Mumps.—This common affection needs little more than careful nursing. A spare diet, rest, and a daily warm bath facilitate recovery. If the diseased parts are very painful, treat as for sore throat. Keep the feet warm. If the breasts or testicles become inflamed, apply ice or alternate hot and cold cloths.

Dysentery.—This disease consists of an inflammation of the large intestine, or colon. In mild cases, the disease is limited to the rectum. The local inflammation is accompanied by general fever, together with the discharge of mucus, with more or less blood. The cause of the disease is sometimes obscure ; improper diet, bad water, foul air, or exposure to wet and cold, during the hot months, may be mentioned as the most common causes of the disease.

In the treatment of this malady, energetic measures should be used to diminish the local inflammation, and to subdue the general fever. This may be done best by the use of fomentations and compresses over the bowels and abdomen, together with the wet-hand rub and wet-sheet pack, as frequently as the severity of the case demands. Great care should be taken to keep the extremities thoroughly warmed. If the head is unnaturally hot, cold applications may be made to it. If spasms occur, great relief may be obtained in an application of ice or very cold water to the head and upper portion of the spine. Local pain may be greatly relieved by the use of warm or cool enemas. Great care should be exercised to keep the patient quiet. His food should be such as will be easily digested, while it is of such a character that it will not be a source of irritation to the mucous membrane.

It is a mistaken notion that fruit is a cause of this disease. It may be occasioned by eating unripe fruit; but the immaturity of the fruit is the cause of the disturbance, being a source of irritation to the intestinal canal on account of its indigestibility. Ripe fruit not only does not occasion dysentery, but some kinds of fruit, as blackberries, raspberries, and grapes, are conducive to recovery when freely used. Fruit is rarely harmful if eaten properly, being taken at meals only, in moderate quantity, and thoroughly masticated.

Diphtheria.—As soon as the first symptoms of the disease appear, begin treatment in a very energetic manner. If the patient is an adult, give him a warm sitz-bath for about twenty minutes. Surround him with blankets during the bath, so as to favor perspiration. The feet should be placed in a hot foot-bath in the meantime, and the head should be frequently wet with cool water. After the bath, quickly sponge the whole body with water a little cooler than that of the bath. Then put the patient to bed and cover him up warm. Keep the feet warm, cool the head by frequent bathing, and sponge the whole body every hour or two with tepid water if the patient is very feverish.

If the patient is a child, a warm pack will be preferable to a sitz-bath. Wring a woolen sheet out of water a little more than blood-warm. Spread it quickly upon the bed, place the patient upon it, and quickly envelop him. Then wrap him snugly with dry blankets, and let him sleep for half an hour if he feels so inclined, as he usually will. Follow the pack by tepid sponging, as directed after the sitz-bath.

After putting the patient to bed, apply, alternately, hot fomentations and cold compresses. Fold a flannel cloth twice, so as to give four thicknesses, wring it out of water as hot as can be borne dry enough so that it will not drip, and apply at once to the throat. After a lapse of three to five minutes, apply a cold compress for the same length of time. Then re-apply the fomentation, and continue to alternate until each has been applied four or five times. Then apply a cool compress, and change it as often as it becomes warm.

In ordinary cases, it will be sufficient to wet the cool compress in the coldest well water that can be obtained; but in cases in which there is great irritation of the throat, snow or pounded ice should be applied, being placed between the folds of the compress.

By all means avoid the use of all of those caustic applications which are so commonly employed in this disease. When white patches appear in the back part of the mouth, touch them every two or three hours with pure lemon juice, using a swab of soft linen or sponge attached to the end of a lead pencil or a small stick.

If the patient is old enough, some relief will be given by using a gargle of water acidulated with vinegar. Another excellent gargle which destroys the vegetable parasites always present in this disease, is a solution of permanganate of potash. The crystals can be obtained of any druggist. Place two or three in a glass of water, and stir until they are dissolved. The fluid should not be taken into the stomach, though no harm will result if a few drops are swallowed.

A very favorite remedy with many physicians, is the inhalation of the vapor of warm vinegar. The vinegar may be heated in a coffee-pot, and inhaled from the nozzle. A plan highly recommended is the inhalation of the vapor which arises when lime is slaked in a vessel. These measures will often give great relief.

The sick-room should be well ventilated, in order to carry away as rapidly as possible the foul germs which result from the disease, and thus prevent their re-absorption into the blood. The diet should be plain and light, though enough should be given to sustain the nutrition of the patient. Oatmeal gruel and mild fruits are usually well received. Milk may be employed when the patient has been accustomed to its use. The same regularity in meals should be observed as in health.

Ague.—Ague, or intermittent fever, is one of the most common of all diseases in malarious districts. It prevails especially in the spring and autumn months. The exciting cause of the disease is a certain poisonous miasm which rises from low lands which are alternately flooded and dried during the warm season.

Bilious or remittent fever is produced by the same cause. These diseases are so common that we need not describe their symptoms.

Prevention.—The following suggestions respecting prevention will be found useful:—

1. Unless compelled by dire necessity to do otherwise, do not live in a malarious district; in other words, seek a residence that is as remote as possible from localities where malaria is known to be produced.

2. If your residence is already fixed in a malarious district, employ every means possible to prevent the reception of the poison into the system and to counteract its effects. Avoid being in the vicinity of the malarious localities during the evening and early morning, since at these times the miasm settles near the ground. Secure, if possible, a dense growth of trees between the source of malaria and the residence; if this is impracticable, plant, every year, in the same place,

a large area of sunflowers, which serve the purpose of destroying the miasm.

3. Keep the system in as free and clear a condition as possible by avoiding such habits and such articles of diet as will impair the integrity of the liver, skin, kidneys, lungs, and other eliminating organs. This will enable the system to eliminate the poison without its occasioning disease.

Treatment.—At the beginning of the disease, give the patient a vapor-bath on the well day, and in case the chill occurs every other day, repeat the treatment on each well day for a week. During the chill, surround the patient with warm blankets, hot bricks, bed-warmers, a jug of hot water, or any other means of imparting artificial heat; but be careful to avoid applying water to the surface of the body, unless it be to the head. Care should be exercised to remove the hot applications as soon as the fever begins to appear. When the fever is at its height, sponge the body with tepid water. The sponging may be repeated at intervals while the fever continues. During the sweating stage, frequently wipe the skin with a soft cloth; and when the sweating ceases, change the patient's clothing after a thorough sponging of the body. If there is a tendency to sweat at night, administer the wet rubbing-sheet at bedtime. If the vapor-bath cannot be given, the wet-sheet pack is a very good substitute.

The diet should be very simple. Oatmeal or

graham gruel, with ripe fruit and dry toast or graham crackers, constitutes an admirable dietary for a person suffering with ague.

In case the chill occurs every day, the vapor-bath or pack should be given in the afternoon or every other day after the paroxysm is past. If the severity of the disease is unabated after this treatment has been thoroughly applied for a week or ten days, it would be well to resort to direct means for breaking the periodicity of the disease. A very efficient means of doing this is to get the patient into a profuse sweat by surrounding him with hot bricks, warm blankets, and other hot applications, twenty minutes before the time for the chill to begin. The patient should be kept very warm for an hour or two, or until all danger of chilling is past. Care should be exercised not to press this means to such a degree as to produce violent congestion of the head. If this plan fails after two or three thorough trials, the use of a very small dose of an antiperiodic medicine will break the chills, and then the patient will make a rapid recovery; but the use of drugs will be very rarely required if treatment is applied efficiently and discreetly. The treatment described has proven successful in a large number of cases. When the cause of the disease is removed from the system, it will usually cease. But in case the paroxysms are not interrupted after the lapse of a reasonable amount of time, a small dose of medicine will do the system less

harm than the prolongation of the disease ; for
the popular theory that it is better to wear out
the disease than to check it in any way, is a great
error.   The long continuance of the disease is exceedingly damaging to the system, while it is in
no way beneficial.   In many instances, consumption, dropsy, and other grave and fatal diseases,
are produced by allowing ague to continue until
the vital forces of the patient are exhausted.

Whooping Cough.—No method of treatment
will *cure* this disease.   The patient gets well of
himself in due time in ordinary cases, if he is
not dosed with sickening, poisonous drugs, teas,
sirups, expectorants, cough mixtures, and emetics.   Good care, plenty of fresh air, a warm bath
three or four times a week, and a plain, nourishing diet, are the best means to secure a speedy
recovery.

Worms.—Various kinds of worms infest the
human body.   Children are particularly liable to
them.   For the small worms which are found in
the rectum, perfect cleanliness, regularity of the
bowels, daily enemas of salt water, and anointing
the anus with sweet-oil, are the best remedies.
Indigestion and constipation are the chief causes.
Tape-worm and the large round worm require
more energetic measures of treatment.   For the
first, the best remedy known is the seed of the
common pumpkin.   Take two ounces of *fresh
seeds*, remove the shells, and beat them to a paste

with an equal quantity of finely pulverized white sugar. Add a little milk or water, and take at one dose after fasting twenty-four hours. After three hours, take a table-spoonful of castor-oil. If this does not dislodge the worm, there probably is none. Many people imagine they have tape-worm when they have not. For a child, the dose should be about one-half that for an adult. The fluid extract of the seeds can be obtained at the stores, the dose of which is half a fluid ounce.

For the round worms, worm seed, *chenopodium*, is one of the best remedies. To a child two or three years old give half a dram of the seed in sirup or honey, night and morning, for three or four days in succession. After the last dose, give a tea-spoonful of castor-oil. Five or ten drops of the oil may be given with sugar in place of the seed.

Constipation.—Torpidity of the large intestine is a condition very common among sedentary people, especially women. It is the result, in part, of eating fine-flour bread and irritating condiments. One of the greatest causes—the chief, perhaps—is neglect to attend promptly to the calls of nature. When the feces are retained in the rectum, they become hard and dry through the absorption of their fluid portion. Thus a considerable part of this foul matter is taken into the system, permeating every fluid and tainting every tissue. The dry, hard residue becomes packed in the intestine, and makes defecation difficult, and is productive of several serious diseases of the bowels and other abdominal organs.

Nothing could be more injurious than the use of purgatives as remedies for this difficulty. No matter under what form or name they are taken, they always aggravate the disease in the end, though they seem to give temporary relief. Besides, these "aperients," "laxatives," "purgative pellets," and "cathartics" are the most potent causes of dyspepsia. To cure the difficulty do this :—

1. Exercise plentifully and regularly in the open air.

2. Eat no bolted flour. Instead, eat wheat meal, or graham flour, oatmeal, rye, barley, crushed wheat, etc. Eat plenty of fruit, sparingly of milk, sugar, and condiments. Discard hot drinks at meals. Knead and percuss the abdomen gently for half an hour each day, or five minutes at a time, and several times a day. By regularity in habits, accustom the bowels to move at a certain hour each day. Secure an action of the bowels at least once each day, if possible, but do not resort to the continued use of the enema to effect it. Drink a glass of cold water half an hour before breakfast, if it does not disagree with the stomach.

Piles.—This malady is simply a result of the preceding one. It usually disappears when its cause is removed. Sometimes, however, the tumors which are formed have to be removed. Ointments seldom do any good. The numerous "infallible cures" advertised, are frauds. Cool bathing of the parts, cleanliness, and the injection of cool water, are among the best remedies.

A horde of quacks are just now infesting the country as " pile doctors." They profess to cure by a secret and painless remedy. They should never be employed.

Cold Feet.—Cold feet are due to deficient circulation. Administer the alternate hot-and-cold foot-bath as directed for chilblains, several times a day, if possible ; at least, twice a day. Wear large, thick boots or shoes, and thick woolen stockings. Keep the feet dry. Exercise. Allow no constriction about the limbs, as garters or elastics. Clothe the upper portions of the limbs warmly. Do not wear rubbers except for a little while at a time when necessary. Electric or galvanic soles are of no use whatever. The feet should be kept perfectly clean, and the stockings should be changed every day, being allowed to air one day, when they may be worn again. Three changes a week are none too many for cleanliness and warmth. Cork soles are useful.

Heart-Burn.—This unpleasant affection has nothing to do with the heart. It is the result of fermentation of the food, which produces irritating acids. These are thrown up into the mouth, producing a burning sensation. A few sips of hot or cold water will commonly give relief.

Sometimes a warm-water emetic is required. Soda and magnesia, which are so often used, are productive of a vast amount of mischief. They never cure, but increase the real disease, and sometimes cause fatal injury to the stomach and intestines.

Crick in the Back.—This curious malady is sometimes relieved as quickly as produced, by stretching the back by bending backward across a log or fence. Hot fomentations, with vigorous rubbing, usually give relief quite readily.

Stitch in the Side.—This difficulty is of the same character as the preceding. Hot applications usually give prompt relief. A tight flannel bandage should be worn about the trunk after the fomentation has been given.

Lumbago.—Alternate hot and cold applications followed by thorough rubbing and percussion are the best local applications. Systematic treatment, and attention to the general health, are also required.

Biliousness.—Every spring the regular doctors, and the quack doctors, and all the drug fraternity, reap a rich harvest from the numerous multitudes who seek to be cured of biliousness by purgatives, alteratives, " blood-purifiers," and " anti-bilious pills." This is one of the great popular delusions upon which charlatans and druggists fatten. The ill feelings which are interpreted to mean too much bile, really mean too much fat pork, too much sugar, too much grease, too much mince-pie, too much cake and preserves, too much fried sausage; in fact, too much of all kinds of food, whether good or bad. April and May bring the penalty of the transgressions of the winter months. Flagrant outrages against Nature in the matter of food and drink are often seemingly borne with impunity during the cold months; but if the same line of conduct is extended into the warmer months, all the symptoms of " biliousness " appear.

The proper cure for " biliousness " is, first, Abstinence for a day or two until Nature can get rid of a little of the grossness which clogs her machinery; second, Avoidance of the cause; third, A few packs, fomentations over the liver, and the daily dry-hand rub, with a wholesome diet. Lemons and other acid fruit seem to have a favorable influence upon this condition of the system.

Bitters are filthy compounds of various nauseous drugs and poisons, and bad whisky. *All* of them contain alcohol. " Temperance Bitters " and " Vinegar Bitters " are no exceptions. Some contain more alcohol and fusel-oil than do brandy, gin, or rum. The various " blood tonics," " purifiers," " invigorators," etc., are of the same character. Their manufacturers are deserving of a.

place in the deepest part of the bottomless pit; for they lay snares for the unwary, making drunkards of the best and most promising men and youth. Their pretensions are all falsehoods, and their testimonials are either fraudulent or the result of bribery. Can bitters purify the blood? Never. As well talk of cleansing a delicate fabric with slime from a cess-pool.

Roots and herbs belong in the same category with the rest. They are not so harmful; however, though equally useless.

Cramps.—Relief is given by the hot or cold douche, hot fomentations, rubbing with cold water, and by pressing the affected muscle against some hard body, or grasping it firmly with the hand. Cramp in the stomach may require an emetic of warm water, with a hot sitz-bath and foot-bath.

Palpitation of the Heart.—Indigestion is the usual cause. It will cease when the cause is removed. It need not be a cause of alarm in ordinary cases. If the patient has had rheumatism he should have his heart examined by a physician.

Indigestion.—Proper food, eaten in proper quantity, and at the proper times, ought to be properly digested. In rare cases, only, it may not be. When it is discovered that an article of food is really injurious to digestion, discard it at once. Eat few kinds at a meal. Avoid eating fruits and vegetables together. Do not drink at meals. Eat slowly. Eat mostly dry food. Do not sleep soon after eating. If the stomach is slow in its action, hot fomentations and gentle kneading soon after eating will promote digestion. Salt and other condiments are often the cause of indigestion.

Sometimes oatmeal gruel, eaten with dry crackers, will be retained and digested when nothing else will be. Other cases will not tolerate any kind of farinaceous food.

A young infant which is for any reason deprived of its natural food, and rejects everything else, will thrive upon a mixture of raw white of egg in water—the white of one egg to a half pint of tepid water. The water should not be hot enough to coagulate the egg. Thoroughly mix, and feed with a spoon.

Softening of the Brain.—So-called softening of the brain is not softening of the brain at all. It is simply congestion of the brain from bad food, bad air, late hours, dissipation, lack of exercise, and sundry other causes. Healthy food, a daily bath, abundant sleep, and plenty of exercise in the open air, will cure nearly every case in a short time.

Consumption.—Is consumption curable? It is, if taken in time. The following hints, if carefully followed, will arrest the disease in its early stages :—

1. Avoid all the causes of the disease, chief among which are breathing air which has been previously breathed, sedentary habits, late hours, and exposure to extremes of temperature.

2. Live in the open air at least seven hours a day. Exercise sufficiently to produce moderate fatigue, but not exhaustion. Walking and horse-back riding are good exercises.

3. Fill the lungs to their utmost capacity several times in succession, every hour of the day at least; and cultivate the habit of deep breathing. Do not strain the lungs by holding the breath long. Keep the shoulders well thrown back.

4. Avoid all kinds of stimulants and stimulating food. Eat the most nourishing kinds of food. The chance for recovery largely depends upon the amount of nutriment which can be well digested and assimilated.

5. Take a thorough tepid sponge-bath, followed by a dry-hand rub, three times a week. The whole body should be thoroughly rubbed with the dry hand each morning.

6. Wear flannel the year round; thick in winter, thin in summer. A silk under-suit is an excellent protective.

7. Avoid every form of cough sirup, balsam, cough mixtures, lozenges, expectorants, etc., etc., no matter how strongly recommended. Cod-liver oil, fat pork, bullock's blood, and similar remedies are as useless as absurd and disgusting.

Be sure to begin in season. A few months' delay has often sacrificed the last chance. "Throw physic to the dogs," obey the laws of Nature, and trust in Nature's God.

Vomiting.—If the patient evidently has something in his stomach which ought not to be there, as indigested food, or something obnoxious which has been swallowed, administer a warm-water emetic to assist in the removal of the cause of the difficulty. If there is no evidence of anything in the stomach which needs expulsion, apply either very cold or very hot cloths over the stomach, place the feet in hot water, and give sips of either *hot* or cold water, or little bits of ice to swallow. The attempt should not be made to check the vomiting unless it is clear that the stomach has been freed from its irritating contents, if this was the cause which induced it at first.

Cough.—Coughing, like vomiting, should be encouraged rather than restrained when there is anything which needs expulsion in that manner. Many consumptives have been suffocated by the sudden stopping of a cough which was merely an effort of Nature to get rid of foul matter in the lungs. If there is no cause for the cough but irritation in the throat, it may be cured, in most cases, by the application of the wet bandage. Wear night and day, and change frequently. If the cough seems to have no sufficient cause, it may be concluded that it is of a purely nervous character. The force of will power is the best remedy. Resolve not to cough, engage the attention with something else, and forget it. This method will sometimes succeed even when there is a little irritation present. Continuous coughing will produce irritation of itself. Frequent sips of cold water, and gargling cold water or a mixture of water and lemon juice, will often relieve a cough when it is due to irritation of the upper part of the windpipe. Wearing the wet bandage about the throat is an excellent remedy.

Do not eat honey, lozenges, loaf-sugar, licorice, hoarhound, cough candy, or anything of the kind. They are worthless as remedies, and do the stomach a vast deal of damage.

Hiccough [hickup].—This troublesome affection is usually caused by a disordered stomach. Get the stomach in good condition, and it will disappear. A few sips of cold water will often relieve it. Perhaps the best remedy is holding the breath and fixing the attention intently upon some object.

Sneezing.—When suddenly seized with a desire to sneeze, place the finger upon the upper lip and press hard. Rubbing the nose vigorously will also suppress the paroxysm when it is desirable to do so. When the affection is caused by disease of the nasal cavity, it will not be so easily controlled. The inhalation of steam, and the warm or cold nasal douche, or gently drawing water into the nose, will frequently give material relief.

Bad Breath.—The chief causes are catarrh, decayed teeth, foul teeth, disordered stomach, and constipation. The remedy is to remove the cause. If there are foul and decaying accumulations in the nose, remove them by syringing the nose with a weak solution of permanganate of potash, common salt, or tepid water. Simply snuffing the fluid gently into the nose is quite effective. The fluid should not be thrown violently into the nose, as injury may result therefrom.

Decayed teeth should be either filled or drawn their presence in the mouth is not only a cause of offense, but is productive of disease of the stomach, besides being a source of impurities which find their way into the blood through the lungs.

Uncleanly teeth are quite certain to decay sooner than those which are kept free from impurities. If the food which adheres to the teeth and lodges between them is allowed to remain, it speedily undergoes putrefaction and becomes very offensive. The teeth should be cleansed with a brush and pure water after each meal, and soon after rising in the morning. Once a day, at least, they should be thoroughly brushed with fine soap and pulverized chalk. Artificial teeth need especial attention. They should be daily washed with fine soap and a solution of carbolic acid and water, in proportion of a teaspoonful of the acid to a pint of soft water. Shake well before using. Do not wear artificial teeth during the night.

A solution of chlorinated soda, which can be procured of any druggist, is a most excellent article for cleansing the mouth and the teeth. It should be used freely.

When disorder of the stomach is the cause, it must be cured, to purify the breath.

If the contents of the bowels are retained, instead of being promptly voided, their fluid portion will be absorbed into the blood with all their noxious and disgusting properties. The characteristic odor can be easily detected in the breath of persons whose bowels are constipated or irregular. Few things are more offensive than the breath of a costive child.

The proper remedies for foul breath from this cause are pointed out under the head, "Constipation." No amount of good looks can atone for a foul breath. Cleanliness and wholesome diet are all that are necessary to remove it. It is a very disgusting thought that the breath may contain what ought to have been voided from the bowels some time before.

Sleeplessness.—Eat an early and light supper of easily digested food; or, better, eat no supper at all. Do not engage in exciting conversation or amusements during the evening. At an early hour prepare to retire, determined to sleep. Just before going to bed, soak the feet for ten minutes in a pail of hot water. Cool the water a little just before taking them out. This will relieve the brain of a little of its surplus blood. Go to bed at peace with all the world, close the eyes, and fix the mind steadily upon some familiar object until sleep comes. Don't allow the mind to wander if possible to prevent it. If unsuccessful, in addition to the above have hot wet cloths applied to the head after going to bed. A dripping-sheet bath just before retiring sometimes affords excellent results. Gently rubbing the temples with the hand, and rubbing the spine from above downward and the feet and limbs in the same direction, have a very soothing effect. The warm full-bath is an excellent soporific.

Ulcers.—Old ulcers on various parts of the body are frequently very offensive as well as painful. To remove the odor emitted by the discharges, wash them thoroughly twice a day in a weak solution of carbolic acid or permanganate of potash. The application will also do something toward healing it. The water-dressing and a strict diet are the best remedial agents.

Chafing.—Fleshy persons and children are often seriously troubled by chafing in hot weather. Daily cleansing of the affected parts with cool water and fine soap, and local tepid bathing, repeated several times a day, will prove the most efficient remedies. Anointing the parts with sweet cream or a little unsalted butter, and applying dry, powdered starch, are useful measures. Cleanliness is the most important remedy.

Canker.—The small white ulcers which sometimes occur in the mouths of both children and adults are commonly known by this name, which really belongs to a much more serious affection. They indicate derangement of the stomach. The proper remedies are, improvement of the digestion, washing the mouth frequently with cold water, and touching the cankers with nitric acid, lunar caustic, or some other caustic application. Various astringent washes are used with some benefit.

Chapped Hands, Feet, and Lips.—Wet, cold, and dirt are the chief causes. The use of poor soap, and imperfectly drying the hands before exposure to cold, are the exciting causes of chapped hands in most cases. To cure, keep them scrupulously clean. Wash them with castile soap and soft water. After wiping them nearly dry, rub them with finely powdered starch.

Washing the hands with water to which a handful of bran or corn meal has been added, is a good remedy.

Another remedy: After thorough washing and

drying, at night, apply glycerine, adding a few drops of soft water, and rubbing in well. Wear gloves during the night.

Sweet cream is another common remedy. Honey is warmly recommended by some. The wet bandage is one of the best of all.

The same remedies are to be used for the lips and feet as for the hands. When fissures, or cracks, occur, keep the edges together by means of adhesive plaster.

Stammering.—Stammering is a real disease. It is sometimes induced, by imitation of others, in those who have no natural impediment of speech. It is rather difficult to cure, but perseverance and firmness will master it. Speak very slowly and deliberately, uttering no sound until the vocal organs are well under control. Open the mouth widely in speaking, speak loudly, and breathe deeply. One of the causes of stammering is attempting to speak with the lungs only partially filled. Stop speaking instantly when the slightest embarrassment is felt, and keep the lungs well filled.

Dandruff.—Cleanse the scalp daily with pure soft water and fine soap, and brush it with a soft brush. Do not use any of the patent nostrums advertised.

Sore Eyes.—Ordinary inflammation of the eyes is greatly relieved by laying upon them one or two thicknesses of linen cloth wet in tepid water. Smarting of the eyes when reading will usually be relieved by moistening them often with water. Never use eye-water or caustic unless under the advice of a skillful oculist.

Near-sightedness.—If the eyes are near-sighted, they should be at once provided with suitable glasses, or they will suffer injury. The glasses should be adapted to the eye by an experienced oculist.

Far-sightedness.—Like the preceding, this disease needs immediate attention, although less injury will result from some neglect in this case.

Baldness.—Cut the hair short, and bathe the head twice a day in cool water, adding considerable friction with a brush of medium stiffness. Keep the feet warm, and maintain good digestion. If the hair follicles are not destroyed, the hair will grow again ; otherwise it will not. The various lotions sold for this purpose are poisonous, and produce diseases which are sometimes fatal.

Itch.—The disease is caused by a parasite which burrows under the skin. The object of treatment is to kill the insect. It is perhaps possible to do this by means of water alone ; but as the only applications necessary are made to the skin only, no harm can result from the careful use of more speedy and effective remedies. Sulphur is the most reliable remedy. Take two ounces of lard, one ounce of sulphur, and one-eighth ounce of powdered sal-ammoniac. Mix well and apply at night after thoroughly washing the affected parts in strong soap-suds. Allow the ointment to remain on over night. Wash it off thoroughly in the morning, and put on clean clothes. Repeat the same treatment three or four times in succession. An ointment of storax and lard, one part of the former to four of the latter, is quite efficient. Perfect cleanliness is essential to successful treatment. The application of oil and lard alone is said to cure by half a dozen applications. Mercurial preparations should be avoided, as they sometimes poison the system.

Lice.—Animal parasites of various kinds which infest the body, abound only when their presence is encouraged by filth. They usually disappear very quickly when absolute cleanliness is preserved. If they do not at once vanish, the application of an ointment made of one part of Scotch snuff to two of lard will speedily destroy them. This ointment is quite poisonous, and should be quickly removed after thorough application.

Warts.—If the wart is small, it may be cured by touching it with the end of a stick which has been dipped in strong acetic acid. The application should be made several times a day until it is destroyed. If large and old, apply nitric acid in the same way. Lunar caustic and caustic potash may also be used.

Corns.—These are excrescences produced by a morbid growth of the skin. They are caused either by friction or by pressure, and are usually the result of wearing a tight and otherwise ill-fitting boot or shoe. Corns are not always produced by tight shoes or boots, being often occasioned by the friction of loosely fitting foot-gear.

There are two varieties of corns, hard and soft. Hard corns are formed upon the outside of the toes; soft corns are produced between the toes.

To cure a corn, the first thing to be done is to soften it. To accomplish this, soak the foot in hot water for one hour every night, and then apply a cloth saturated with a strong solution of saleratus. Continue this treatment for three or four days; then remove the corn with a thin, sharp-bladed knife, carefully working the instrument between the corn and the healthy skin beneath. If the whole corn has been removed, all that now remains to be done is to protect the part from pressure. This may be very easily accomplished by placing over it a piece of soft buckskin, in which an opening has been made of the exact size of the corn, which should be placed exactly over the seat of the disease. By this simple means, the diseased surface will be wholly protected from pressure. Any tendency to harden will be prevented by keeping the buckskin saturated with sweet-oil. This simple treatment, if thoroughly applied, will rarely fail to cure any corn.

Bunions.—These originate in the same way as corns, and require somewhat similar treatment. Soaking the feet in hot water when they are inflamed, and bathing with cool water at other times, gives great relief. If there is much thickening of the skin, apply a caustic, as nitrate of silver, or lunar caustic. When the black surface comes off, apply the caustic again. Wearing a piece of soft buckskin, as directed for corns, to prevent pressure, is a useful remedy.

Boils.—The application of heat and cold, alternately, will sometimes disperse a boil in the early stage. When it becomes painful, apply hot fomentations frequently, with the wet compress during the intervals, or apply continuously a soft poultice. The wet compress covered with oil-silk has the same effect as the poultice. The kind of poultice is quite immaterial, if it be un-irritating, for its only valuable properties are warmth and moisture.

When the boil is ripe, that is, when a little white vesicle appears near the surface, its cure may be hastened by lancing with a sharp knife. The discharge may be encouraged by gentle pressure ; but squeezing boils is a very harmful process, and greatly retards their cure. If they do not discharge freely after opening, poultice or apply fomentations. Applications for the treatment of boils should be made to the surrounding tissues as well as to the boil itself, to be effective.

A carbuncle is simply a large boil. A sty is a small one upon the eyelid. Treatment for each is the same as for ordinary boils.

It is a mistaken notion that the purulent matters discharged from boils are concentrated impurities which previously existed in the blood. The pus itself is made up of the white blood corpuscles, the most precious part of the blood. The discharge contains impurities, but most of them are the result of the death of the tissues which have suffered in the inflammation. It is yet an undeniable fact that many persons experience an improvement in health after having several boils, whatever may be the explanation. The contents of a boil are very poisonous to the system when absorbed into the blood.

Stone-Bruise.—This disease, usually the result of accident, is of a nature similar to felon. The intense pain often present is relieved by placing the part in very cold water. It may be treated nearly like a felon.

Felon.—The real disease is an abscess formed beneath the periosteum, or skin of the bone. It may sometimes be dispersed by the application of turpentine or other strong irritants, or by holding the finger in strong lye as hot as can be borne for half an hour, several times a day. Keeping the hand constantly in ice-cold water gives great relief, and sometimes prevents the further progress of the disease if employed in time. Relief is also afforded by the cold douche, arm-bath, and wearing the cold compress upon the arm and hand. When the disease is manifestly settled, the quickest remedy is found in lancing the finger to the bone. Warm fomentations and poultices may afterward be applied, to encourage the discharge.

Hang-Nail.—If the toe is greatly inflamed, it should be placed in a warm foot-bath, an hour at a time, three times a day. During the intervals, it should be covered with a poultice made of bread and milk, linseed, or slippery elm. By this means, the inflammation and tenderness will be greatly reduced. The next step in treatment should be to scrape the center of the nail with a sharp knife until it becomes as thin as possible without exposing the flesh. Then slightly elevate the outer edge of the hang-nail for the purpose, and place underneath it delicate pledgets of cotton. If the nail penetrates the flesh so deeply as to make this impossible, it may be necessary to remove a very small portion by splitting it off with a sharp knife. A still better way is to crowd underneath the diseased portion of the nail delicate filaments of floss-silk, drawing in one portion after another until the nail is elevated out of the tender flesh. The nail may be still farther elevated by the employment of the same means, while the poultices are continued, till a complete and permanent cure is effected.

Diseases of Women.—The declining health and strength of American women has come to be a very common observation. Very few young ladies of the present day can compare with their grandmothers of the last generation in powers of physical endurance. Physicians generally acknowledge that at least three-fourths of their practice is derived from diseases of women. The causes of this general and notable decline are well worth consideration. We will briefly hint at a few.

Fashionable Dress.—No one cause has done more to undermine woman's physical health than her devotion to dress. Whatever fashion dictated, she has felt in duty bound to follow, no matter if in so doing she committed the grossest violations of the laws of health. In thus doing, she has compelled her poor body to undergo the most inhuman tortures. She has heaped upon her sensitive, nervous head, a cruel load of artificial hair; nearly choked herself to death with belts and corsets, and squeezed her vital organs into most unnatural shapes; contorted her tender feet into misshapen masses with tight shoes and high heels; and disturbed her whole vital economy by surrounding her vital organs with a super-abundance of clothing while suffering her limbs to go almost unclad, no matter how cold and damp the weather. With such abuse is it strange that she complains of headaches, lung troubles, weak back, and general debility?

Sedentary Habits are another prolific cause of woman's decline. Confinement within doors, without a proper amount of physical exercise, results in deficient development of the muscular system, and various weaknesses follow which render her feeble and inefficient. Too much novel reading, piano thrumming, parlor lounging, and day-dreaming are ruining the constitutions of thousands of the young ladies of the present day.

Late Hours.—Fashionable dissipation at any time is bad enough ; but when continued to a late hour of the night, or even until early morning, when the system is exhausted for want of rest, it becomes doubly enervating. Sleep is Nature's opportunity for repairing the wastes which occur during the hours of wakefulness. The nervous system wears out faster than any other part of the body ; hence it suffers more severely than any other part when deprived of proper opportunity for repair. Is it any wonder, then, that so many ladies are nervous and hysterical, and constantly complaining of headaches, neuralgias, and weak nerves ?

Bad Diet.—Improper dress, with deficient exercise and late hours, with the usual accompaniments of dancing and feasting, so enervate the system as to create a demand for artificial stimulation, in the form of strong tea and coffee, mustard, pepper, spices, animal food, and all sorts of highly seasoned dishes. The certain result of this abuse of the digestive organs is dyspepsia in some one of its myriad forms. Torpidity of the liver and skin are accompanying evils which may properly be traced to the same cause. The loss of that clearness and brilliancy of complexion which exist only in health, leads to the use of cosmetics of various sorts, which, in many cases, still further undermine the health and injure the skin.

Sexual Sins.—One of the most potent though usually obscure causes of woman's physical decline, is that referred to in the heading of this paragraph. Transgressions of Nature's laws in this regard are attended with results the most fearful that humanity can suffer. Sexual excesses, for which she is usually only in small degree responsible, not only occasion their own sad results, but lead to the perpetration of such horrible crimes against Nature as prevention of conception, and fœticide or abortion. Thousands of women have by some form of sexual transgression brought upon themselves diseases and weaknesses which entail life-long suffering. These evils are becoming so prevalent that unless checked they threaten to exterminate the race.

Too Much Drugging.—Last, but not least, in the list of enemies to woman's health, we mention drugs. Medicines of this class undoubtedly have their legitimate place ; but they are subject to great abuse. The general tendency of most of the other causes mentioned is to produce obstinate constipation of the bowels. For this evil a remedy is sought in laxatives of various sorts, after-dinner pills, and purgatives. These give temporary relief, only to exaggerate the difficulty which they are expected to remove. Tonics are demanded to support the waning strength, which is not replenished by proper rest and well-digested food. Nervines and opiates are required to quiet the weak and irritable nervous system. Chloral and morphia become indispensable to procure sleep. Headaches and neuralgias necessitate fresh doses of narcotic drugs. Hysterical attacks call for antispasmodics. General debility is an indication for stimulants, while torpor of the liver, skin, and system generally, suggests the need of alteratives. Thus the life becomes a daily round of dosing. One after another various drugs lose their effect, and are replaced by others more powerful. Meanwhile the system grows daily weaker, more torpid, and more diseased.

Such trifling with Nature is in the highest degree reprehensible, and will prove fatal to the strongest constitution. Drugs never cure such maladies. No remedy is of any value which does not reach the causes of the diseased conditions to be removed. If the women of America value health, if they covet physical strength, if they aspire after the endurance of their grandmothers, let them abandon the ruinous habits which are dragging them down, and enervating their mental and physical forces. Let them shake off the shackles of fashion and conventionality, and conform to the God-implanted laws which govern their sensitive bodies.

Care of the Sick.—Every physician knows that in the majority of cases much more depends upon the care which the patient receives from his nurse, than from himself. A careless nurse has often turned the scale, which hung nearly evenly balanced between life and death, adverse to recovery. The following are some of the more essential matters which demand attention, though nothing can supply the native tact and grace which are necessary to make a good nurse :—

1. Secure a constant supply of pure air from out-of-doors. It is not sufficient to open a door leading into another room. Cold air may be very impure. Be careful to exclude the air from the kitchen and wash-room as perfectly as possible.

2. Admit the light and sunshine freely. Direct sunlight is sometimes unpleasant to the patient; then shade the windows with white curtains, which will admit the light. In a few diseases it may be necessary to keep the patient in a darkened room for a few days.

3. Maintain equable temperature. More fire is needed in the morning and evening than at noon. Regulate the heat by a thermometer hung near the bed. The mercury should never be above 70°. Old people especially need attention in this particular. A fall of a few degrees in temperature is often fatal to them. Avoid draughts.

4. The linen of the patient and his bedding should be changed every day at least. Daily washing will not be demanded in all cases, but the clothing should hang for several hours near a heated stove to air and dry.

5. Food for sick people should always be simply and neatly prepared. Light food is usually the best. Condiments should be very sparingly added, if at all. Oatmeal gruel is one of the best articles of food for sick persons. Fruit may be freely allowed if of good quality and ripe. Beef tea and broth will not sustain life. A dog starved sooner on a diet of beef tea than he would have done with nothing at all. Give food regularly, as in health. Continual dosing with milk or any other food is harmful.

6. The patient himself should be kept scrupulously clean. The whole body should be washed several times a week at least. The mouth and teeth should be daily cleansed.

7. All discharges should be kept in covered vessels, and should be removed from the room at the earliest moment possible.

8. The sick chamber should be made pleasant by tasteful arrangement of its contents, by flowers, simple pictures, etc. Frequent change in the aspect of the room is desirable.

9. The patient should never be kept in a state of expectancy. When a promise is made him, fulfill it promptly.

10. Whispering or low talking in the sick-room

or adjoining rooms will arouse the patient's fears unnecessarily. Avoid it.

11. Hasty movements in the sick-room are always annoying to a patient. A calm, deliberate air on the part of the nurse inspires confidence.

12. Arrangements for the night should be made before the patient becomes sleepy, so that he may not be disturbed. Otherwise, the movements necessary in making preparations for the night may cause him to become so restless that sleep will be impossible.

13. All avoidable noises should be prevented. Creaking doors, squeaking boots or shoes, a swinging blind or a flapping curtain, are intolerable to the sensitive ears of invalids. Coal should never be poured from the scuttle upon the fire. Bring it into the room in small parcels wrapped in damp paper. These can be laid upon the fire noiselessly.

14. If the patient can sleep, let him sleep. Never think of waking a sick person out of a sound sleep. Refreshing sleep will do him more good than all the medicines and baths in the world.

15. The covering of the patient in bed should be several light, porous blankets, rather than one or two heavy ones.

16. Strangers and visitors should be prohibited from entering the sick-room of a feeble patient. Visiting will often determine a fatal issue of the disease.

17. Water kept in a sick-room should be often

changed. Never drink that which has been in the room more than a few minutes.

18. Always wear a cheerful face. Do not look solemn and anxious, even though the case is grave.

19. Never annoy the patient by questions or too much conversation.

20. Always recollect that Nature must cure. All you can do is to make the conditions as favorable as possible.

Signs of Real Death.—It has sometimes happened that people have been buried alive when they were seemingly dead. Such a sad mistake can be prevented by the following tests :—

1. The loss of sensibility and warmth, and cessation of the pulse and the breathing, are the signs which at first indicate death ; but these are not always reliable.

2. Rigidity of the muscles is another better evidence, but this is not wholly decisive ; yet if the muscles remain soft after death, interment should be delayed.

3. The most reliable sign of death, perhaps the only decisive one, is putrefaction. This usually begins first in the lower part of the abdomen.

4. Another test of some value in doubtful cases is tying a cord tightly around a finger. If death has taken place, the color will remain unchanged. If the heart still beats, the end of the finger will become of a deeper color.

5. The application of a hot iron or other caustic appliance will not produce a blister on a corpse.

Accidents and Emergencies.

THE injuries resulting from accidents usually demand instantaneous action. A little delay or confusion, or misdirected effort, in a case of severe burning, drowning, or hemorrhage, will often sacrifice a human life. The following simple directions should be carefully studied so that they can readily be made available at any moment :—

Drowning and Suffocation.—The chief remedy to be used in all cases is *artificial respiration*. There are several methods which are very serviceable. The following, which is the most approved method for restoring drowned persons, we copy from a publication issued by the Michigan State Board of Health, the Secretary of which, Dr. H. B. Baker, has kindly furnished us with cuts for illustration :—

TREATMENT OF THE DROWNED.—"Two things to be done : 1. Restore breathing ; 2. Restore animal heat.

"RULE 1.—*Remove all obstructions to breathing.* Instantly loosen or cut apart all neck and waist bands ; turn the patient on his face, with the head down hill ; stand astride the hips with your face toward his head, and, locking your fingers together under his belly, raise the body as high as you can without lifting the forehead off

the ground (Fig. 1), and give the body a smart
jerk to remove mucus from the throat and water
from the windpipe; hold the body suspended long
enough to slowly count *one, two, three, four, five,*
repeating the jerk more gently two or three times.

Fig1.

" RULE 2.—Place the patient on the ground,
face downward, and, maintaining all the while your
position astride the body, grasp the points of the
shoulders by the clothing, or, if the body is naked,
thrust your fingers into the armpits, clasping your
thumbs over the points of the shoulders, and *raise
the chest as high as you can* (Fig. 2) without lifting the head quite off the ground, and hold it long
enough to *slowly* count one two, three. Replace
him on the ground, with his forehead on his flexed
arm, the neck straightened out, and the mouth and
nose free. Place your elbows against your knees,
and your hands upon the sides of his chest (Fig. 3)

over the lower ribs, and press downward and in-ward with increasing force long enough to slowly

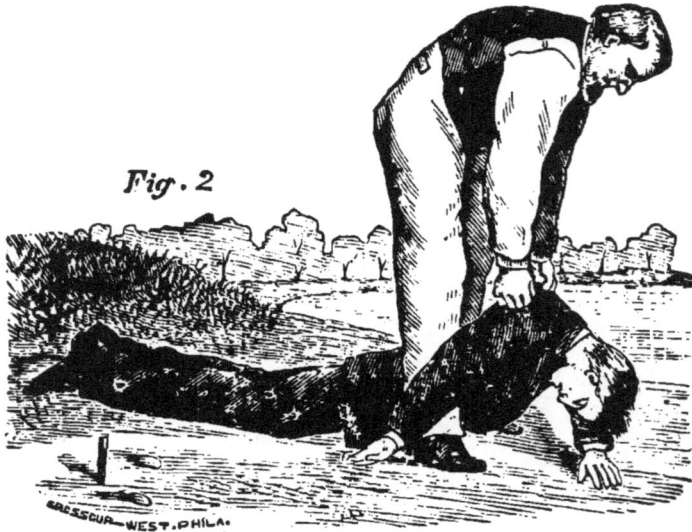

Fig. 2

count one, two.   Then suddenly let go, grasp the

Fig. 3.

shoulders as before and raise the chest (Fig. 2);
then press upon the ribs, etc. (Fig. 3).   These al-

ternate movements should be repeated ten to fifteen times a minute for an hour at least, unless breathing is restored sooner. Use the same regularity as in natural breathing.

"RULE 3.—After breathing has commenced, RESTORE THE ANIMAL HEAT. Wrap him in warm blankets, apply bottles of hot water, hot bricks, or anything to restore heat. *Warm the head nearly as fast as the body, lest convulsions come on.* Rubbing the body with warm cloths or the hand, and slapping the fleshy parts may assist to restore warmth, and the breathing also. If the patient can SURELY swallow, give hot coffee, tea, milk, or a little hot sling. Give spirits sparingly, lest they produce depression. Place the patient in a warm bed, and give him plenty of fresh air; keep him quiet.

"*Avoid Delay.* A MOMENT may turn the scale for life or death. Dry ground, shelter, warmth, stimulants, etc., at this moment are nothing— ARTIFICIAL BREATHING IS EVERYTHING—is the ONE REMEDY—all others are secondary.

"*Do not stop to remove wet clothing before efforts are made to restore breathing.* Precious time is wasted, and the patient may be fatally chilled by exposure of the naked body, even in summer. Give all your attention and effort to restore breathing by forcing air into, and out of, the lungs. If the breathing has just ceased, a smart slap on the face, or a vigorous twist of the hair will sometimes start it again, and may be tried incidentally, as may, also, pressing the finger on the root of the tongue.

"Before natural breathing is fully restored, do not let the patient lie on his back unless some person holds the tongue forward. The tongue by falling back may close the windpipe, and cause fatal choking.

"If several persons are present, one may hold the head steady, keeping the neck nearly straight; others may remove wet clothing, replacing at once clothing which is dry and warm; they may also chafe the limbs, and thus promote the circulation.

"*Prevent friends from crowding around the patient and excluding fresh air;* also from trying to give stimulants before the patient can swallow. The first causes suffocation; the second, fatal choking.

"*Do not give up too soon.* You are working for life. Any time within two hours you may be on the very threshold of success without there being any sign of it."

MARSHALL HALL'S READY METHOD.—This famous method consists, briefly, in laying the patient with his face downward, his arms folded beneath his forehead, and then slowly rolling him upon his side, restoring him again to his former position. By this means, the chest is alternately compressed and expanded, thus imitating the movements of respiration. This method has been variously modified.

SYLVESTER'S METHOD.—This method, which has been proposed more recently, is highly recommended by many physicians. Raise the arms from the sides until they meet above the head; then bring them slowly back to the sides again, pressing them against the sides of the chest. Repeat this sixteen or eighteen times a minute. It is a very efficient means when skillfully applied.

Upon submersion in the water, the epiglottis, a little valve at the top of the windpipe, closes, shutting out the water from the lungs. After a time, the muscles relax, and the valve opens. Water then enters the lungs. After this occurs, there is no longer any possible chance for recovery; but as there is no ready means for determining accurately the condition of the lungs, every effort should be made to resuscitate the patient by the means already described. The length of time a person can live under water will depend very much upon the amount of pure air in his lungs at the time of submergence.

Poisonous Gases.—Carbonic acid (more properly carbon di-oxide) is the most common cause of suffocation. Chlorine gas, illuminating gas, the vapor of burning sulphur, ether, and nitrous oxide, or laughing gas, with other poisonous gases, produce death in the same way, though some of them are active irritants in addition.

Carbonic acid is heavier than air, and, in consequence, it accumulates in old wells, caves, deep valleys, and other low places. It is formed in mines in large quantities, at times, and is known to miners as " choke damp." It is also formed in the vats of breweries by fermentation. In the burning of limestone it is also produced in enormous quantities. When the kilns are opened, it sometimes pours out so rapidly as to suffocate the workmen before they can escape. Miners are often destroyed by a sudden gust of "choke damp."

Old wells should never be entered without first testing the air at the bottom. Do this by lowering a burning candle. If it is extinguished, or burns feebly, carbonic acid is present, and descent would be extremely perilous. If it burns brightly, no fears need be entertained. If gas is found to be present, it can be dislodged by throwing into the well burning fagots or paper. Old cellars and cisterns are sometimes dangerous on the same account; they may be tested in the same way.

Upon the inhalation of the first breath of carbonic acid, the person usually falls, and thus remains exposed to the poisonous effects of the gas. Under such circumstances, speedy and well-directed efforts are necessary to prevent death.

In a burning building, the purest air is near the floor, as the smoke containing the carbonic acid is hotter than the air when first formed, and rises. In escaping from a burning building, it is sometimes advantageous to go upon all-fours so as to breathe the best air.

Charcoal burning in a room in an open vessel will produce large quantities of carbonic-acid gas in a short time. In France, suicide is often committed by this means.

Illuminating Gas often escapes into sleeping-rooms through leakage of the gas pipes, or by reason of failure to completely shut off the supply to the burner upon extinguishing the flame.

People unaccustomed to the use of gas are sometimes so thoughtless as to blow out the flame as they would that of a lamp or candle, leaving the gas to find unobstructed entrance. Many lives have been lost in this way.

Hanging is another means by which the supply of air to the lungs is cut off, causing asphyxia. A red line around the neck is usually indicative of this manner of producing suffocation.

The remedies in all cases of suffocation are essentially the same. Remove the patient from the cause, or, *vice versa*, as quickly as possible. Draw the tongue forward, clear the mouth, dash cold water upon the face and chest, rub the body vigorously, and apply artificial respiration. If chlorine has been accidentally breathed, inhale, as quickly as possible, ammonia gas.

Choking.—When a particle of food, or any other body, becomes lodged in the throat, go upon all-fours, and cough. If it is not expelled, the patient should be seized by the heels and suspended head downward, while his back is percussed by another person. If the body can be seen by drawing the tongue well forward, seize it with a pair of forceps, or a hook made by bending the end of a wire or a hair-pin which has been straightened.

Sometimes it may be elevated from its position by
means of a spoon handle.   If it is out of sight, and
all efforts to expel it are unavailing, press it down
with the finger or a smooth rod with a rounded end,
throwing the head back as far as possible while doing so.   A body which has lodged part way down
the œsophagus, may sometimes be pressed down into
the stomach by pressing hard upon each side of the
neck close to the windpipe.

Lightning Stroke.—Dr. Fothergill remarks as
follows on this subject :—

"Persons struck by lightning are not always
dead when they appear to be so.   There are few
recoveries from this state, because no means are
tried to restore the sufferer.   In the tropics there
are many instances of persons, struck down by
lightning, recovering after a heavy thunder shower;
and it would appear that cold affusion to the body
has a decided action in such cases.   The injured
cannot be harmed by the free use of cold water,
and if only an occasional recovery took place, it
would be well worth the pains bestowed.   The persons so injured should have cold water poured or
even dashed freely over them."

Artificial respiration should also be employed.

Sun-Stroke.—Carry the patient at once to a
cool, shady place, remove his clothing, and dash
cold water upon his body, especially the head and
chest.   Rubbing the spine with ice is an excellent
remedy.   Continue the cold application until the

unnatural heat is materially decreased. Artificial respiration should be practiced at the same time. No stimulants should be given to the patient.

Hemorrhage.—If an artery is wounded, the blood will flow in jets, sometimes being thrown several feet, and will be of a bright red color. If the wounded vessel is a vein, the blood will be of a dark color, and will flow in a steady stream. Slight hemorrhage will be easily controlled by pressure over a little pad of folded linen applied directly to the wound

When large vessels are injured, greater care is necessary. If the vessel is an artery, apply the pressure between the wound and the heart. If it is a vein, apply the pressure upon the opposite side.

The application of cold, by means of cloths wet in iced water, snow, or pounded ice, is a verv effective means of stopping hemorrhage.

In case the hand, forearm, or foot is severely wounded, it should be elevated above the rest of the body and bound in towels in which pounded ice is folded. Hemorrhage from the end of a finger or toe may be stopped by the application of pressure to the sides.

When a very large artery of the arm or leg is wounded, resulting in hemorrhage which cannot be quickly controlled by any of the means mentioned, proceed as follows :—

Take a handkerchief or a strip of cloth of sufficient length to reach around the limb. Tie a large knot in the center. Apply the knot just over the course of the wounded vessel, above the wound. Now pass a stout ruler or rod beneath the bandage upon the opposite side from the knot. Twist it around so as to tighten the bandage and thus compress the artery beneath the knot. Increase the compression until the hemorrhage is controlled. A tight bandage of this kind should not be retained too long, as it may destroy the life of the parts below. Its object is to control the hemorrhage only until the wounded vessel can be secured and tied by a surgeon or other competent person.

Bleeding from wounds of the scalp is easily controlled by pressure upon the seat of injury.

Nose-Bleed.—Remove all constrictions from the throat, so that the return of blood from the head will be unobstructed. Hold the head erect for the same reason. Inclining it forward encourages the hemorrhage. Twist the corner of a handkerchief or piece of old linen and press it tightly into the bleeding nostril. Hold it in place until the bleeding ceases, unless it passes backward into the throat, when other measures will be required. Blowing the nose, and bathing it in water, increase the hemorrhage rather than check it.

Pressure upon the facial artery upon the side on which the hemorrhage occurs, will sometimes check it. Apply firm pressure upon the notch on the lower border of the lower jaw just in front of the angle.

When the bleeding has once stopped, do not disturb the clot that has formed in the nose, as it may be induced again by so doing. In very severe cases the posterior opening from the nasal cavity into the mouth will require plugging; surgical assistance will be required for this.

Hemorrhage from the nose is seldom fatal. When scattered upon the floor or clothing, a few ounces of blood look like a quart. A very few spoonfuls will color a large quantity of water very red.

Bleeding from Lungs.—Blood which is expectorated by coughing often comes from the throat or nasal cavity, trickling down into the air passages and being coughed out. This is nearly always of a dark color, and is commonly clotted. Blood which comes from the lungs is of a bright red color, and is frothy from the admixture of air. The amount of blood lost is much less than usually thought, and is seldom the cause of death.

Keep the patient as quiet as possible, with his head elevated a little. Instruct him to restrain his cough as much as he can, and to avoid all violent efforts at coughing. Make cold applications to the chest and spine, and hot to the feet and limbs. For applying cold to the chest, rubber ice-bags are very convenient, as they do not wet the clothing. When they are not at hand, employ compresses of snow or pounded ice large enough to cover the entire chest, or the affected side if the exact origin of the hemorrhage is known. Salt and other drugs are often employed; but it is exceedingly doubtful whether they are of much value, since they pass at once to the stomach, not entering the lungs at all.

Cuts.—Cuts should be dressed in such a way that the severed edges may unite properly. Firm clots of blood lying in the wound should be carefully removed, with any other foreign body. If the bleeding has ceased, the edges may be brought together and secured by stitches or adhesive straps, according to the size and position of the wound. Small wounds sometimes require only that the edges be thus brought together to stop the bleeding. The strips of plaster used should be narrow, and there should be narrow spaces left between them, to allow room for the escape of the discharge, should any occur.

Care should be taken not to close a wound when vessels of any size have been ruptured without either ligating the bleeding vessel or closing it by torsion. Much injury has often resulted from a neglect of this rule.

If the end of a finger or toe has been accidentally cut off, it should be at once replaced, even though it was entirely severed. Being kept in place, it will be quite likely to adhere and prevent an ugly scar. If the severed piece is frozen or badly bruised, the attempt will be useless.

Dressing for Wounds.—As a dressing to be applied to all wounds, nothing is equal to water. While swollen and painful, cold applications should be made by means of thin compresses, which should be changed every few minutes. After the pain and inflammation have subsided, apply thin compresses kept constantly wet with tepid water. In some cases submersion of the part in water is serviceable.

The various " pain-killers," liniments, and washes have no healing virtue whatever. Opium and arnica relieve the pain only by paralyzing the nerves. They simply hide the condition of the wound from the patient. Both are poisons which retard healing.

Bruises.—Apply as quickly as possible a hot fomentation. Renew the application every five minutes for an hour or two. Apply afterward the tepid compress. This will prevent soreness, and much of the swelling and discoloration which would otherwise result. This is the way to treat a black eye, a broken nose, or a foot which has been pierced by a rusty nail.

How to Cure a Sprain.—A sprain is an injury to a joint, produced by straining or lacerating one or more of the ligaments connected with a joint. The first thing to be done after the receipt of the injury is to apply hot fomentations to the injured joint ; and the sooner the better. After applying hot fomentations for one or two hours or longer, if the pain continues, apply cold compresses and keep the joint entirely at rest. When there is much swelling, alternating it with cold pouring, continued for an hour at a time, will often give

great relief.   Rest is one of the most essential feat-
ures of treatment, since the injured ligaments cannot be repaired while disturbed by the motion of
the joint.   Cases are numerous in which an injury
which was at first a slight sprain, has resulted in
the total loss of the use of the limb, from neglect to
give the joint the required rest while Nature was
effecting a repair.   The various liniments which
have a reputation for the cure of sprains are useful
only as a means of inducing the patient to rub and
manipulate the joint.   Rubbing is a very useful
means of treatment, especially if the limb is con-
siderably swollen.   Violent manipulation of the
joint should be carefully avoided, as it would only
serve to increase inflammation.

Fractures and Dislocations.—These accidents usually require the attention of a skillful surgeon, who should be called at once.

Burns and Scalds.—If a person's clothes
catch on fire, wrap about him at once a blanket,
cloak, rug, or similar article, bringing it tight about
the neck to protect the head and face.   Remove the
burned clothing as quickly as possible, and apply
wet linen cloths to the burned surfaces.   Change
every five minutes, applying another cloth instantly after one is removed.   (For further treatment
see page 82.)

To burns produced by lye, caustic potash, or
other alkalies, apply vinegar or some other weak
acid as quickly as possible.   To a burn produced by

an acid, apply an alkali, as soda, ashes, or simple earth.

Freezing.—In cases of freezing, the great danger is in thawing out too quickly, the result of which is inflammation and death of the frosted parts; or, in milder cases, chilblain. Keep the patient away from the fire. Place him in a cool room, and rub the frozen parts with snow or cold wet cloths until the circulation is re-established. If the patient is apparently dead, artificial respiration should be practiced as long as there is a particle of hope of recovery; and the effort should not be abandoned for several hours.

Those who are exposed to severe cold should remember that one of the symptoms of freezing is an uncontrollable desire to sleep. Resist it.

Bite of Mad Dog.—Remove the clothing from the part at once, and apply suction to the wound with the mouth. As quickly as possible, remove the injured flesh with a sharp knife or destroy it with an iron at white heat, afterward applying the water-dressing or a poultice.

Few persons that are bitten by rabid animals ever have the disease. Hydrophobia is more common among dogs in the winter than in the summer, contrary to the common supposition. The skunk or polecat is liable to the same disease. Its bite is more dangerous than that of the dog.

Rattlesnake Bite.—Destroy the poison virus in the same manner as described in the preceding article. As with the bites of mad dogs, few of those bitten are poisoned, and fewer still fatally so. Artificial respiration and rubbing the spine with ice have been highly recommended. Whisky is entirely worthless as an antidote. It does more harm than good when administered.

Insect Stings.—The pain caused by the sting of an insect is the result of an acid poison injected into the tissues. The first thing to be done is to press the tube of a small key firmly on the wound, moving the key from side to side to favor the expulsion of the sting with its accompanying poison. The sting, if left in the wound by the insect, should be carefully extracted, as it will greatly increase the local irritation. The poison of the virus being acid, common sense points to the alkalies as the proper antidote. Among the most easily procured remedies may be mentioned soft soap, liquor of ammonia (spirits of hartshorn), smelling salts, washing soda, quicklime made into a paste with water, limewater, the juice of an onion, bruised dock leaves, tomato juice, wood-ashes, and carbonate of soda. A solution of borax in proportion of one ounce to a pint of water is also a most excellent remedy.

The same remedies should be applied to the bites of gnats, mosquitoes, spiders, fleas, and other insects.

Dirt in the Eye.—Particles of dirt or other foreign bodies in the eye should be removed at once. If the object is upon the visible portion of the eyeball, remove it with the corner of a handkerchief. If concealed beneath the lid, roll the lid over upon a pencil or turn it outward with the finger, and remove the speck in the same way. Dirt beneath the upper eyelid can often be removed by drawing it outward and downward over the under lid. Then press it upon the under lid and open the eye. Blowing the nose while the eye is closed will assist in the removal of small particles of dirt. Particles of iron which have become imbedded in the tissue of the eye may be loosened and removed by a needle mounted in the end of a pencil ; but such an instrument must be used with extreme care.

Lime in the Eye.—Lime is a powerfully caustic alkali, and in numerous instances a small quantity thrown into the eye has resulted in total destruction of sight. A strong solution of sugar or diluted vinegar should be applied as quickly as possible after the accident, in case a particle has been thrown into the eye. While the lotion is in preparation, the eye should be thoroughly washed.

Foreign Bodies in the Ear.—Never use a sharp instrument about the ear in any way. Insects can generally be dislodged very speedily by dropping into the ear a little oil or warm water. Solid bodies, like peas, beans, or pieces of stone, can usually be removed by the diligent application of warm water and soap by means of a syringe. The head should be inclined to one side, so that the object may readily drop out. If this is unsuccessful after thorough trial, use a loop of fine wire or horsehair, a small scoop, or a pair of delicate forceps. Hardened ear-wax should be softened by warm water and soap, and then removed with great care by means of the scoop.

Foreign Bodies in the Nose.—Blow through the nose with as much force as possible, at the same time closing the mouth and the unobstructed nostril. Sneezing will sometimes expel the cause of obstruction. A loop of wire or a blunt hook may be successfully used; but care must be taken to avoid crowding the object farther in. If it is not tightly imbedded, it may be driven out by making the water from a syringe pass up the unobstructed nostril and out at the one containing the foreign body.

Another plan is to blow the patient's nose for him by closing the empty nostril with the finger, and then blowing suddenly and strongly into the mouth. The glottis closes spasmodically, and the whole force of the breath goes to expel the button or bean, which commonly flies out at the first effort. This plan has the great advantages of exciting no terror in children, and of being capable of being at once employed, before delay has given rise to swelling and impaction.

Chimney on Fire.—Throw into the stove, or upon the coals in the fire-place, a handful of salt or sulphur. Close the stove-draught, or hold a board or blanket before the fire-place.

What to do in Poisoning.—Give an emetic at once, which may consist simply of tepid water in large quantities, or the same with the addition of mustard or common salt. After drinking several cupfuls, tickle the throat with the finger or a feather. Continue taking a cupful every two or three minutes until vomiting occurs. Individual poisons require special remedies. The following lists comprise the most common poisons and their antidotes :—

Vegetable Poisons.—Opium, Morphia, Camphor, Aconite, Laudanum, Paregoric, Strychnia, Tobacco, Lobelia, Arnica, and other vegetable poisons require the emetic and the application of a stomach-pump if possible. Milk and mucilaginous drinks should be given freely after thorough vomiting. Artificial respiration should be employed in poisoning by strychnia and opium. The cold douche is also excellent in poisoning by the latter drug. Keep the patient awake, if possible, by making him walk about.

Acids.—Sulphuric (oil of vitriol), Nitric (aqua fortis), Hydrochloric (muriatic), and Oxalic Acids are the more common. Drink largely of water at once. Acids are neutralized by alkalies. Calcined magnesia is the best antidote. Chalk (powdered), whiting, lime, weak lye, and strong soap-suds are the best substitutes. Something must be done quickly in case of poisoning by acids.

Mineral Poisons.—For Corrosive Sublimate, White Precipitate, Red Precipitate, and Vermilion, take the whites of several eggs in a quart of tepid water. Soap-suds thickened a little with wheat flour is the best substitute for eggs. No other emetic is necessary.

Arsenic, Cobalt (fly powder), Ratsbane, Paris Green.

and other compounds containing Arsenic, should be expelled by vomiting as soon as possible. Then administer quite large doses of calcined magnesia.

Acetate of Lead, White Lead, Litharge, and Saltpeter require an emetic followed by oil or mucilage.

For Lunar Caustic (nitrate of silver), administer half a table-spoonful of salt in a pint of water.

The antidote for Matches or Phosphorus is calcined magnesia, followed by soothing fluids.

Antidotes for Verdigris and Blue Vitriol (sulphate of copper), are eggs, milk, and soda.

Alkalies.—The most common which are sources of poisoning are Ammonia, Potash, Soda, Pearlash, Lye (from wood-ashes), and Salts of Tartar. Drink copiously of weak vinegar or lemon juice. Afterward take some mucilaginous drink, or oil.

Alcoholic Poisoning.—A man found " dead drunk " should be treated like any other case of narcotic poisoning, as from opium.

Chronic Poisoning by Lead, Opium, Tobacco, or any other drug which has been received into the system for a long time, requires, first, that the cause be wholly removed at once ; second, attention to the general health. In the case of Opium and Tobacco, the disuse of the drugs is attended with a good deal of unpleasant feeling on the part of the patient. He feels as though he will certainly die. His fears are groundless. He is in much less danger of dying than before.

Poisonous Candies and Food.—The paints used in the manufacture of candies are poisonous, and often sicken those who eat the candies, sometimes fatally in the case of children.

Fish and meat, either fresh or canned, are frequently sources of poisoning. Decayed fruit or other food, shell-fish, and mushrooms are often productive of injury in the same way. Such cases should be treated on the general principles relating to poisoning.

Soda-Water.—The water nearly always contains lead. The sirups are most wretched imitations of natural flavors, and are made from such things as old cheese, tar, and mineral acids.

Dangerous Kerosene.—The kerosene oil sold or used in the majority of our cities is almost as dangerous a commodity as gunpowder or nitro-glycerine. Millions of dollars' worth of property has been destroyed, and hundreds of lives have been sacrificed, by the use of cheap illuminating oil. Crude kerosene contains benzine, naphtha, and other highly volatile and explosive compounds. These dangerous agents should be wholly removed by the refiner in preparing the oil for use ; but the manufacturer finds it to his pecuniary advantage to allow them to remain in the oil in greater or lesser proportions. This kind of oil will burn at a much lower temperature than that which is pure, and it is to this fact that its dangerous properties are due, since it is thereby rendered explosive when used in the ordinary kerosene lamp.

It is very important to be able to distinguish dangerous oil from that which may be used without danger. The following is an excellent method for testing oil :—

Place upon the stove a pan or tin pail containing water. Float in this vessel a deep saucer or small, deep cup containing a portion of the oil to be tested. Place in the oil a thermometer, and observe the gradual increase of temperature. When the temperature reaches 70° or 80°, bring a burning match or taper near to the surface of the oil. If a flash is produced, the article is highly dangerous. Continue the observations as the temperature rises, and if a flash is observed at the temperature less than 140°, the oil is utterly unfit for use, and should not be employed for illuminating purposes.

The lower the temperature at which the flash occurs, the greater the danger.

The State Legislature of Michigan has passed an act prohibiting the use or sale of kerosene oil which will flash below 140°.

Hydropathic Appliances.

WATER, applied in the various modes in which it may be, is one of the most potent of remedies. Wrongly applied, it may be productive of great harm. The following are a few general rules which should always govern its use :—

1. Never bathe when exhausted or within three hours after eating, unless the bath be confined to a very small portion of the body.

2. Never bathe when cooling off after profuse sweating, as reaction will then often be deficient.

3. Always wet the head before taking any form of bath, to prevent determination of blood to the head.

4. If the bath be a warm one, always conclude it with an application of water which is a few degrees cooler than the bodily temperature.

5. Be careful to thoroughly dry the patient after his bath, rubbing vigorously to prevent chilling.

6. The most favorable time for taking a bath is between the hours of ten and twelve in the forenoon.

7. The temperature of the room should be at about 80° or 85°.

8. Baths should usually be of a temperature which will be the most agreeable to the patient. Cold baths are seldom required. Too much hot bathing is debilitating.

The following are brief descriptions of the more important baths applicable in the home treatment of disease :—

Sponge-Bath.—This bath consists in rubbing the whole body with a sponge or towel wet in water of an agreeable temperature; is most useful for a general ablution.

Sitz-Bath.—A tub made especially for the purpose, or a common wash-tub, may be· employed. Place in the vessel sufficient water to cover the hips and lower part of the abdomen. The patient or an attendant should rub and knead the abdomen during the bath. The water should be of a temperature ranging from 85° to 98°, according to the condition of the patient. Cover the patient during the bath.

Wet-Sheet Pack.—Spread two or three comfortables upon a bed or mattress. Spread over the whole a woolen sheet. Wring out of water of the desired temperature a linen or cotton sheet. Spread it quickly upon the bed, and let the patient immediately lie down in the middle. Then quickly envelop him in the wet sheet, wrapping him snugly from head to foot. Then cover him with the comfortables, and let him remain quiet as long as required. Elevate the head a little, and use care to have the feet warm. Half-packs may be taken in a similar manner. confining the application to the trunk of the body.

Fomentations.—Wring out of water as hot as can well be borne, a folded flannel cloth, and apply it quickly to the part to be treated. Cover with a dry cloth, and change once in five minutes.

Pail-Douche.—This consists in pouring water over the shoulders of the patient with a pail. It is often employed to tone up the surface after a hot bath.

Chest-Wrapper.—The wrapper should be made of coarse cloth, and should be shaped so as to fit the chest. Apply it after wringing just sufficiently to prevent dripping. Cover with a light, dry flannel wrapper. Change three or four times a day.

Half-Bath.—For this bath is required a vessel of sufficient size to allow the patient to sit upright with his limbs extended. Enough water to cover the limbs, thighs, and lower part of the abdomen, is necessary. During the bath, the attendant should rub vigorously the limbs, back, chest, and abdomen of the patient.

Compresses.—Apply wet cloths in the same manner as in fomentations, wetting them in either cold, cool, or tepid water, according to the effect desired.

Rubbing-Wet-Sheet.—This bath consists in enveloping the patient in a wet sheet, and rubbing him briskly with the hand outside the sheet.

Hot Applications.—Besides fomentations, heat may be applied in several other ways. Bottles filled with hot water, hot bricks or stones wrapped in papers or cloths, hot cloths, bags filled with hot sand, salt, or corn meal, and rubber bags filled with hot water, are convenient methods of applying dry heat.

Moisture and heat may be applied in a variety of ways also. Instead of wringing cloths out of hot water, put them into a steamer for a few minutes. This saves the trouble of wringing them. When there is no water hot, and a fomentation is wanted quickly, wring a cloth out of cold water, spread it between the folds of a newspaper, and lay the paper upon the top of the stove, or press it against the side. In a minute it will be hot. Wrap stones or bricks in a moist cloth. Poultices of various sorts answer the same purpose.

All hot applications should be renewed every few minutes until the desired effect is obtained.

Vapor-Bath.—Place the patient in a chair which has a wooden bottom, beneath which place a pail half filled with water. Surround the patient completely, chair and all, with a woolen blanket, leaving only his head visible; even this may be covered a little while at a time in cases of neuralgia, if desired. Add other blankets sufficient for warmth. Now raise the blankets a little, behind, and place in the pail a stone or brick which has been heated sufficiently hot to hiss when it touches the water. Do not drop it into the water at once, but let it in gradually. As this becomes cool, add another in the same way. The bath should not usually be continued more than twenty minutes. Wash off quickly with tepid water upon coming out of the bath. The head should be wet from the first.

Hot-Air Bath.—Prepare the patient in the [same manner as directed for the vapor-bath. Instead of the pail of water, place beneath the chair a cup containing a small quantity of alcohol. Wet the head well, and then light the alcohol. Wash with tepid water after the bath, and be careful to avoid taking cold.

Enemas.—An enema is a small portion of water thrown into the rectum by means of a syringe. The water may be either cool, tepid, or warm, as occasion may require.

Inunction.—Pure olive oil, or fresh butter, may be used, but vaseline, a fine unguent which can be procured of the druggist, is the best. After giving the patient a short bath of some kind, to cleanse the skin, dry him carefully, and then apply with the hand a very small quantity of the oil or unguent. Rub in very thoroughly, with much kneading and friction. Conclude by carefully wiping the skin with a soft flannel to remove all superfluous oil.

Useful Hints and Recipes.

Soap to Remove Grease Spots.—Take equal parts of soft soap and fuller's-earth. After beating well together, form into cakes. Moisten the spot, and rub the soap upon it. Allow it to dry, then rub it well with warm water, rinse, and dry.

To Remove Grease from Silk.—Grease may be removed from silk and other delicate fabrics, thus: Upon a smooth surface spread a woolen cloth. Lay upon this the silk with the right side down. Over the grease spot lay a piece of coarse brown paper. Place upon this a flat-iron sufficiently hot to just scorch the paper. A very few seconds will suffice. Remove the flat-iron and paper and rub the spot briskly with a piece of paper. If this is not quite successful, apply a little powdered chalk or magnesia to the spot, under the brown paper, before applying the flat-iron.

To Restore Color.—When the color has been destroyed by acids, apply a little ammonia (hartshorn). The restoration will be the more perfect, the more recent the application of the acid.

To Remove Stains from the Hands.—For fruit stains, apply a solution of oxalic acid, and wash quickly. Another way: Light a sulphur match and clasp the hand about it while the sulphur is burning.

To Remove Paint from Cloth.—Apply spirits of turpentine with a sponge. After an hour or two, rub the spot as in washing, and the paint will crumble off.

Calcimining Fluid.—The following is well recommended for walls: White glue, 1 lb.; white zinc, 10 lbs.; Paris white, 5 lbs. Soak the glue over night in

3 qts. of water. Add an equal quantity of water, and heat on a water bath until the glue is dissolved. Put the two powders into another vessel. Pour on hot water while stirring, until of the consistency of thick milk. Mix the two liquids thoroughly, and apply to the walls with a whitewash brush.

To Remove Mildew.—Wet the linen, apply soap to the spot, and then apply fuller's-earth or salt and lemon juice to both sides. Air for a few hours. Or, soap the spot, and then apply finely powdered chalk, rubbing it in very thoroughly.

Chloride of lime will remove mildew. Dissolve one ounce in two quarts of water. Steep the linen in the solution all day.

To Remove Paint from Wood.—Apply to it a strong solution of oxalic acid, when it will easily crumble off. It may be removed from glass or metal in the same way.

Cements for Glass and China.—1. Mix thoroughly an ounce of pure white lead in oil with ten grains of finely powdered acetate of lead. Apply at once, and allow the mended article to dry two weeks before it is used.

2. Rub old cheese to a fine thick paste with a little water. Add one-fourth pulverized lime. One of the best cements for glass, porcelain, stone, and wood.

3. Burn oyster shells, pulverize fine, and mix to a thick paste with white of egg. Apply at once to the edges of the glass. Secure them tightly together until dry. Freshly burned lime will do, but is not so good. The cement must be made when used.

4. Soak Russian isinglass in water over night, to soften. Then heat until it is dissolved.

Liquid Glue.—Fill a bottle two-thirds full of common glue. Fill the bottle with whisky. It will dissolve in a few days, when it will be ready for use. Must be kept tightly corked.

Cements for Iron.—Take equal parts of sulphur and white lead, with about a sixth of borax, mixing them so as to form a homogeneous mass. When about to apply it, wet it with sulphuric acid and place a thin layer of it between the two pieces of iron, which should then be pressed together. In a week it will be perfectly solid, and no traces of the cement will be apparent. This cement is said to be so strong that it will resist the blows of a sledge hammer.

2. Mix to a paste with vinegar 5 parts clay, 1 part salt, and 15 parts of iron filings. It will stand heat.

Cement for Stone-ware.—To a cold solution of alum add plaster of Paris sufficient to make a rather thick paste. Use at once. It sets rather slowly, but is an excellent cement for mending broken crockery, eventually becoming as hard as stone.

How to Remove Rust from Clothing.—Oxalic acid will take rust or any other stain out of white goods. Dissolve a small quantity in boiling water and dip the spots in. The acid can be got at any drug store. Another way is to saturate the spots with lemon juice and spread the cloth in the sun; if it don't take out all the rust the first time, repeat the application.

To Clean Looking-Glasses.—Wash with a sponge wet in lukewarm soap-suds. Wipe dry, and rub with buckskin or a newspaper and finely powdered chalk. Polish windows in the same way.

To Cleanse the Hair.—Rub thoroughly into the hair the white of an egg. Wash with soft water until the egg is entirely removed. This leaves the hair soft and pliable. Never use alkalies or coarse soap on the hair.

Fire-Proof Paint for Roofs. Slack stone-lime in a covered vessel. Take 6 qts. of the slacked lime, after it has been passed through a sieve, add 1 qt. of salt, and 1 gal. of water. Boil and skim. Add ½ lb. powdered alum, ¼ lb. pulverized copperas. Then slowly add 6 ozs.

of powdered potash. Finish by the addition of 2 lbs. of fine sand. Apply to the roof with a brush. It may be colored as desired; is very durable, and stops leaks in the roof.

Lotion for Fetid Perspiration.—Permanganate of potash, 1 dr., dissolved in half a pint of water. Wash the part twice a day. A wash of weak vinegar is quite as efficient in some cases.

Cement for Wood.—Dissolve a pound of glue in three pints of water. Add 2 ozs. of powdered chalk and ½ oz. of borax.

To Preserve Steel from Rust.—Cover the surface with finely powdered unslacked lime. The surface may first be smeared with melted tallow before the lime is sprinkled on, to cause it to adhere.

To Clean Leather.—Leather which is uncolored may be easily cleaned by wiping it with a sponge moistened in a solution of oxalic acid.

To Make Cloth Uninflammable.—1. To a quart of boiling water add 1 lb chloride of calcium, and 1 lb. acetate of lime. Moisten the fabric in the solution, and dry.
2. Moisten the goods in a solution of phosphate of ammonia. Dry with a warm flat-iron.

Ink Stains.—Apply a solution of oxalic acid to the spot, and wash quickly. If a reddish stain is left, apply a solution of chloride of lime.

Removing Fruit Stains.—Pour boiling water upon the stained spot, and it will usually disappear. This should be done before the spot has been wet with anything else.

Coal-Tar for Fence-Posts.—Coal-tar is an excellent preservative for fence-posts, if properly used. It should not be used alone, since it contains acids which are destructive to the wood; but when combined with quicklime it becomes a most effective preservative. Mix half a bushel of quicklime with a few gallons of water, and thoroughly mingle it with a barrel of coal-tar. Apply freely to the portion of the post which is to be in contact with the earth.

Carron-Oil.—Mix equal parts of linseed-oil and lime-water. Shake well. Good for burns.

To Determine the Capacity of a Round Cistern.— Square the average diameter. Multiply three-fourths of this amount by the height. This will give the number of cubic feet. Divide by four, and the result will be the number of barrels which the cistern will hold. The following table will be found useful for reference :—

Contents of a round cistern for every foot in depth of

4 feet in diameter, . . .	93 gallons.
6 " " " . . .	212 "
8 " " "	375 "
10 " " " . .	588 "
12 " " " . .	848 "
16 " " " . . .	1500 "

To Ascertain the Weight of Hay.—It is often necessary for the farmer to estimate a quantity of hay without the aid of scales. Here is a convenient method : Find the cubic contents of the stack in feet. Divide by 27, to find the number of cubic yards. A cubic yard of old hay in the stack weighs about 200 lbs. New hay weighs about two-thirds as much. The weight is readily ascertained by multiplying the number of cubic yards by the weight of a single yard.

Remedy for Mosquitoes.—Pour kerosene into the stagnant pools where mosquitoes are generated. This will prevent their hatching, and will be found to be the most efficient means of getting rid of them.

Adhesive Cloth.—Dissolve five ounces of gum arabic in a half pint of hot water. Add glycerine in sufficient quantity to make the mixture about the thickness of sirup. Stretch on a frame, fine muslin or linen cloth. Apply a coat of thin mucilage. When this is nearly dry, apply the mixture as rapidly as possible. Several coats will usually be required.

To Take off Paint.—Slack three pounds of good lime in water. Mix with one pound of pearlash to the thickness of paint. Lay it on the paint to be removed with an old brush and allow it to remain twelve or fifteen hours, after which the paint can be scraped off very easily.

Plant Wash.—An excellent wash for shrubs and large plants is made by dissolving two ounces of pulverized borax in one quart of hot water. Apply with a brush to the stems. It will destroy the green fungi which sometimes infest plants.

Starch Polish.—1. Melt together at a gentle heat 1 oz. white wax and 2 ozs. spermaceti. Add a piece the size of a pea to starch sufficient for a dozen pieces.

2. Dissolve 2 ozs. of gum arabic in a pint of hot water; bottle and cork. Add a table-spoonful to each pint of starch.

Paste.—Mix 8 parts of flour and 1 part of powdered alum with a little water. Beat out the lumps, and pour on boiling water until of the proper consistency, stirring briskly all the time. This is more adhesive than ordinary paste, and will last much longer.

To Color Black.—For 10 lbs. of goods, dissolve and boil ¾ lb. blue vitriol in sufficient water to cover the goods. Dip them three quarters of an hour, airing often. Then remove to another dye made by boiling 6 lbs. of logwood in a sufficient quantity of water for half an hour. Dip three quarters of an hour, air, and then dip three quarters of an hour more. Wash in strong suds.

To Color Scarlet.—For two lbs. of goods, mix together and dissolve in sufficient water 1 oz. cream of tartar; 1

oz. cochineal, well pulverized; 5 ozs. muriate of tin. Boil the dye and place the goods in it. Work them briskly for a quarter of an hour, after which boil an hour and a half, stirring slowly while boiling. Wash in clear soft water, and dry in the shade.

To Color Blue.—For five lbs. of goods, dissolve ¾ lb. alum, ½ lb. cream tartar. Boil the goods in the solution for half an hour. Throw them into warm water.

To Color Green.—1. First, color yellow by soaking the goods in a solution made by steeping together 1 lb. fustic and ¼ lb. alum for 1 lb. of the goods. Remove the chips and add indigo, a table-spoonful at a time, until the desired color is obtained.

2. Make a yellow dye with yellow-oak and hickory bark in equal quantities. Add indigo until the desired shade is obtained.

Tooth Powder.—To make a most excellent and perfectly harmless tooth powder, mix eight parts of precipitated chalk with one part of calcined magnesia. Flavor with a few drops of wintergreen or cinnamon oil if desired. Apply this to the teeth twice a day with a soft brush and pure soft water, or water and fine soap, and they will always glisten like ivory.

Washing Fluid.—Boil together 1 lb. of sal-soda, ½ lb. of stone-lime, and 5 qts. of water, stirring while boiling. Let it settle, pour off the clear fluid, and preserve for use in a stone jug.

Soak the clothes an hour or two in warm suds. Wring out, and soap the most dirty places. Add a tea-cupful of the fluid to a boiler half full of boiling water, and then add the clothes. It will save half the labor of washing, and will not injure the texture of the goods.

To Get Rid of Rats.—Scatter potash freely in their holes and runways. It will make their feet and mouths sore, and they will leave in disgust. Several varieties of traps are quite successful in catching them. Poisoning is not a very good plan, as the dead bodies of those which happen to eat the poison are usually left in some unobserved or inaccessible place, where they undergo decay.

Liquid Bluing.—Pulverized Prussian-blue, 1 oz.; oxalic acid, pulverized, ½ oz.; dissolve in 1 qt. of soft water. Use one or two table-spoonfuls to a tub, according to its size. Will not speck.

To Kill Ants.—Pour into their nests hot water, lime-water, or a strong solution of alum. A little turpentine applied about the sugar barrel will drive every ant away from it.

Wash for the Teeth.—1. Dissolve 1 dr. of carbolic acid with 2 ozs. of alcohol. Add this to half a pint of water. Use freely with a tooth-brush. Is excellent as an application to cleanse artificial teeth.

2. Dissolve 1 dr. of permanganate of potash or soda in ½ pt. of water. Place in a bottle and cork tightly.

Black Ink.—2 ozs. extract of logwood; 2 drs. bichromate of potash; 1 dr. prussiate of potash. Dissolve the logwood in 2 qts. of soft water, soaking it over night and then boiling. Then add the bichromate and prussiate of potash after pulverizing. When the solution is complete, filter, and it will be ready for use. This is a very excellent ink.

Red Ink.—Mix 1 dr. of aqua ammonia, a bit of gum arabic as large as a hazel nut, equal parts of No. 40 and No. 6 carmine, as much as will dissolve, and 7 drs. of soft water. It will be ready for use in a day or two.

Indelible Ink.—Dissolve ½ sc. of nitrate of silver in a teaspoonful of aqua ammonia. In 2½ teaspoonfuls of soft water dissolve 1 sc. of gum arabic. When the gum arabic is dissolved, add an equal weight of carbonate of soda. Mix the two solutions and boil in a bottle placed in a basin of boiling water. When it becomes black, it is ready for use.

Soft Soap.—Cut fine 4 lbs. white soap in bars, and dissolve in 4 gals. of soft water by heating.   Add 1 lb. of sal-soda, dissolve and mix

Bug Poison.—Mix 2 ozs. alcohol, ¼ oz. camphor, ½ oz. turpentine, and 1 dr. corrosive sublimate.   Apply to infested places with a feather.

To Etch on Metal.—Mix two parts of muriatic acid with one of nitric acid.   Cover the surface of the metal with melted wax.   When the wax is cold, write or draw upon it the desired name or design, with a sharp-pointed instrument.   Be careful to remove the wax quite down to the surface of the metal.   Apply the acid with a brush or feather, carefully filling the outlines of the design.   In a few minutes wash the acids away with water, and wipe the surface with oil after removing the wax.

Borax Wash.—Dissolve 1 oz. of borax in 5 qts. of water.   This is a good cleansing wash for the hands, and is also an excellent washing fluid.   Many use it for the hair.   It is rather severe for the latter purpose.

Plant-Lice.—Shower the plant with a solution of carbolic acid in water, a dram to a pint; or fumigate with tobacco smoke.

Mending Tin-Ware.—Every house-keeper can save many dollars by mending her own pans, dippers, and basins.   If a hole in a basin is to be stopped, scrape the inside of the basin just around the hole until it is bright. Dip the end of a little wooden rod in the fluid, and rub it upon the scraped surface.   Now place a small bit of solder over the hole, and heat the under surface over a candle flame until the solder melts.   In a minute it cools, and the hole is stopped.

To Dry Boots.—Fill them with oats at night after removing them from the feet.   Set them in a warm room. In the morning, shake out the oats and the boots will be found to be dry, and will not be shrunken and stiff as they would otherwise have been.

Blue Ink.—Dissolve sufficient indigo in soft water to give the desired color; is very good for ordinary use but will fade.

Soldering Fluid.—Dissolve in 1 oz. of muriatic acid as much zinc as possible. Add ½ dr. of sal-ammoniac.

Solder for Tin.—Melt together 5 ozs. of lead and 3½ ozs. of tin.

Solder for Lead.—Melt together 1 oz. tin and 2 ozs. lead.

Freezing Mixture.—The following are a few of the best known means for producing artificial cold :—

1. Mix 4 ozs. of saltpeter and 4 ozs. of sal-ammoniac, each finely pulverized, with half a pint of water.

2. Mix equal parts of powdered nitrate of ammonium, carbonate of sodium, and water.

3. Mix quickly together two parts of finely powdered ice or snow with one part of salt. This mixture will produce a temperature of 4° below zero.

The article to be frozen should be surrounded by the freezing mixture as quickly as possible after the preparation of the latter. When it is a liquid, it may be contained in a bottle, which can be broken after the freezing is effected, if necessary.

To Extract Grease Stains from Wall-Paper.—Oil marks can be taken from the paper on drawing-room walls, and marks where people have rested their heads, by mixing pipeclay with water to the consistency of cream, laying it on the spot and letting it remain till the following day, when it may be easily removed with a pen-knife or brush.

Disinfecting Fluid.—The following is a recipe for one of the cheapest and most efficient disinfecting fluids known :—

Heat two pounds of copperas in an old kettle for half an hour, stirring frequently. When cold, dissolve the copperas in two gallons of water. Add two ounces of carbolic acid, and mix well together. A pint of this solution poured into the kitchen sink every few days will keep it free from odors. It will also be found very useful for disinfecting the discharges of typhoid-fever patients, for which purpose a little should be kept in the vessel constantly. Even privy vaults can be kept in a comparatively harmless condition by the liberal use of this solution.

To Remove Potato Sprouts.—Place the potatoes in barrels, about one bushel in each barrel. Tilt the barrel upon its edge, and roll it about with sufficient vigor to give the potatoes a thorough shaking. By this means the sprouts will be broken off; and by the repetition of the process once in a week or two, the potatoes may be kept free from young shoots.

To Make Cloth Water-Proof.—Into a bucket of soft water put ½ lb. sugar of lead and ½ lb. powdered alum. Stir occasionally until the solution becomes clear, then pour it off into another bucket, and immerse the garment in it. Allow the garment to remain in the solution twenty-four hours. Scotch tweed is the best material for a water-proof cloak.

There are several other methods: 1. Moisten the cloth on the wrong side with a weak solution of isinglass. When this is dry, apply a solution of nut-galls. 2 Moisten with a strong solution of soap, and then with a solution of alum. 3. Spread the cloth on a smooth surface with the wrong side up. Rub it with pure bees-wax until it is gray. Pass a hot iron over it, and brush it while still warm.

How to Make a Filter.—Take a large flower pot or earthen vessel, make a hole one-half inch in diameter in the bottom, and insert in it a sponge. Place in the bottom of the vessel a number of clean stones of sizes varying from that of an egg to an apple. Place upon this a layer of much smaller stones and coarse gravel. Then fill the jar within two inches of the top, with equal parts of pulverized charcoal and sharp sand, well mixed. Place loosely over the top of the jar, white flannel cloth, allowing it to form a hollow in the middle of the jar, into which the water can be poured. Secure the edges by tying a stout cord around the outside of the jar. By keeping a suitable vessel under the filter thus made, and supplying rain-water when needed, very pure water can be obtained. It can be kept in a cool place in the summer. It will require to be renewed occasionally by exchanging the old sand and charcoal for fresh. The flannel and sponge must be frequently cleansed.

Durable Whitewash.—Slack, with abundance of hot water, half a bushel of lime, stirring briskly meanwhile. When completely slacked, add sufficient water to dissolve. To this add two pounds of sulphate of zinc (white vitriol) and one pound of common salt. The last-named ingredients cause the wash to harden, and prevent cracking. If a cream color is desired, add yellow ochre. For stone color, add raw umber and lampblack.

Cleaning Bottles.—Small shot, pebbles, or broken charcoal, placed in a dirty bottle and shaken about with warm water and soap, will remove almost any kind of dirt. Charcoal is especially serviceable in removing unpleasant odors from bottles.

To Keep Water Cool.—Ice is almost universally depended upon as a means of cooling drinking water in summer. The free use of iced water is harmful. By making use of the following means, the water may be kept sufficiently cool to answer all the real demands of nature; in fact it may be kept nearly at freezing temperature :—

Place between two sheets of thick brown paper, a layer of cotton half an inch thick. Fasten the ends of the sheets together so as to form a roll. Sew in a bottom made of similar material, making it nearly air-tight, if possible. Fill a pitcher with cold water, and cover it with the cylindrical box by inverting it over the pitcher.

If the box is kept constantly wet with water, evaporation will go on so rapidly that the water in the pitcher will be kept very cool for a long time.

Water may also be kept cool by placing it in jugs and wrapping them with wet cloths.

Preserving Grapes.—Pick carefully the later kinds of grapes. Select such bunches as are perfect, rejecting all upon which there are any bruised grapes, or from which a grape has fallen. Spread them upon shelves in a cool place for a week or two. Then pack them in boxes in sawdust which has recently been thoroughly dried in an oven. Bran which has been well dried may also be used. Dry cotton is employed by some. Keep in a cool place. In this way, grapes may be kept until long after New Year's with ease.

Another method still more efficient is to select perfect bunches, as already directed, and dip the broken end of the stem of each bunch in melted sealing-wax. The bunches may then be wrapped in tissue paper and placed in layers, or hung in a cool place, or they may be packed in sawdust.

Japanese Method of Cooking Rice.—Put the rice into a kettle with just enough water to prevent its burning to the bottom. Put on a close-fitting cover, and set over a moderate fire. The rice is thus steamed, rather than boiled. When it is nearly done, remove the cover and allow the surplus steam and moisture to escape.

Rice cooked in this manner turns out a mass of snow-white kernels, each separate from the other, and as much superior to the soggy mass usually produced, as a fine mealy potato is to one which is water-soaked.

Beef Tea.—Although not to be recommended as an article of diet, beef tea is frequently a valuable article of food for the sick, especially if properly made. Pound and cut the beef until it is reduced to a pulp, then place it in a dish and cover it with a very little cold water. Allow it to steep gently for two hours, then strain off the juice, and it is ready for use. Some tastes will require the addition of a minute quantity of salt. One-half pound of beef is required for a pint of tea. A very excellent plan is to place the beef in a bottle with the water, and then place the bottle in a kettle of cold water, which should be gradually brought to the boiling point.

The Bushel.—Weight is the only proper standard for the bushel, being the only accurate one. The following are the weights per bushel for the most common articles of commercial exchange :—

	Pounds.		Pounds.
Wheat,	60	Dried apples,	57
Shelled corn,	56	Dried peaches,	28
Ear corn,	70	Coarse salt,	50
Oats,	32	Fine salt,	56
Rye,	56	Lime (unslacked),	80
Buckwheat,	50	Irish potatoes,	60
Barley,	48	Sweet potatoes,	55
Corn-meal,	48	White beans,	60
Bran,	20	Castor beans,	46
Clover seed,	60	Beets,	50
Timothy seed,	45	Parsnips,	44
Flax seed	56	Carrots,	50
Hemp seed,	44	Onions,	50
Blue-grass seed,	14	Turnips.	42
Green apples,	57	Rutabagas,	56

Uses for Ashes.—There is no more valuable fertilizer than common wood-ashes ; but in order that they should retain their virtue, they should be kept under cover. Ashes which have been leached have very little value.

Ashes are also valuable for disinfecting purposes. They are even better than dry earth for deodorizing animal excreta. A privy may be kept entirely free from foul odors by their liberal use. When employed in this way, their disinfecting and fertilizing properties are both utilized.

Another use for ashes which the farmers would do well to take advantage of, is due to their power of destroying various kinds of insects. Turnips and cabbages may be protected from the ravages of various insects which feed upon them, by sprinkling upon and around them a few ashes daily, for a short time. A practical farmer also asserts that unleached wood-ashes will permanently destroy potato bugs, if sprinkled upon the vines while they are moist with dew, or immediately after a rain.

Cheap Paint for Barns and Sheds.—A very cheap paint may be made by mixing unslaked water-lime with milk to the proper consistency. It adheres well to wood, brick, mortar, or stone when no oil or paint has been previously applied. It makes a very durable coating, and its cheapness leaves nothing to be desired. Skim milk is even better than new milk. Many farmers could greatly improve the appearance of their premises by covering with this simple paint their barns, sheds, fences, and out-buildings.

To Preserve Shoes and Boots.—Do not expose them to extreme heat by warming them too near the stove. The smell of leather indicates that they are already injured. The wearing of rubbers is very injurious to leather. Rubbers should be worn as little as possible, and should be removed from the feet as soon as their use is not absolutely necessary. Every two or three weeks, wash the leather with a cloth moistened in warm water, and when nearly dry, apply a warm mixture of equal parts of neat's foot oil and tallow. Ordinary blacking contains oil of vitriol, and this removes the oil from the leather and causes it to become dry and brittle.

INDEX.

www.ingramcontent.com/pod-product-compliance
Lightning Source LLC
Chambersburg PA
CBHW030849270326
41928CB00008B/1283